T0136979

Smart Innovation, Systems and Technologies

Volume 71

Series editors

Robert James Howlett, Bournemouth University and KES International,
Shoreham-by-sea, UK
e-mail: rjhowlett@kesinternational.org

Lakhmi C. Jain, University of Canberra, Canberra, Australia;
Bournemouth University, UK;
KES International, UK
e-mails: jainlc2002@yahoo.co.uk; Lakhmi.Jain@canberra.edu.au

About this Series

The Smart Innovation, Systems and Technologies book series encompasses the topics of knowledge, intelligence, innovation and sustainability. The aim of the series is to make available a platform for the publication of books on all aspects of single and multi-disciplinary research on these themes in order to make the latest results available in a readily-accessible form. Volumes on interdisciplinary research combining two or more of these areas is particularly sought.

The series covers systems and paradigms that employ knowledge and intelligence in a broad sense. Its scope is systems having embedded knowledge and intelligence, which may be applied to the solution of world problems in industry, the environment and the community. It also focusses on the knowledge-transfer methodologies and innovation strategies employed to make this happen effectively. The combination of intelligent systems tools and a broad range of applications introduces a need for a synergy of disciplines from science, technology, business and the humanities. The series will include conference proceedings, edited collections, monographs, handbooks, reference books, and other relevant types of book in areas of science and technology where smart systems and technologies can offer innovative solutions.

High quality content is an essential feature for all book proposals accepted for the series. It is expected that editors of all accepted volumes will ensure that contributions are subjected to an appropriate level of reviewing process and adhere to KES quality principles.

More information about this series at http://www.springer.com/series/8767

Yen-Wei Chen · Satoshi Tanaka
Robert J. Howlett · Lakhmi C. Jain
Editors

Innovation in Medicine and Healthcare 2017

Proceedings of the 5th KES International Conference on Innovation in Medicine and Healthcare (KES-InMed 2017)

 Springer

Editors
Yen-Wei Chen
College of Information Science
  and Engineering
Ritsumeikan University
Kusatsu, Shiga
Japan

Satoshi Tanaka
Ritsumeikan University
Kusatsu, Shiga
Japan

Robert J. Howlett
KES International
Bournemouth University
Poole, Dorset
UK

Lakhmi C. Jain
University of Canberra
Canberra, ACT
Australia

ISSN 2190-3018   ISSN 2190-3026   (electronic)
Smart Innovation, Systems and Technologies
ISBN 978-3-319-86616-1   ISBN 978-3-319-59397-5   (eBook)
DOI 10.1007/978-3-319-59397-5

Printed on acid-free paper

This Springer imprint is published by Springer Nature
The registered company is Springer International Publishing AG
The registered company address is: Gewerbestrasse 11, 6330 Cham, Switzerland

Preface

The 5th KES International Conference on Innovation in Medicine and Healthcare (InMed-17) was held on June 21–23, 2017, in Algarve, Portugal, organized by KES International.

The InMed-17 is the 5th edition of the InMed series of conferences. The 1st, the 2nd, the 3rd, and the 4th InMed conferences were held in Italy, Spain, Japan, and Spain, respectively. The conference focuses on major trends and innovations in modern intelligent systems applied to medicine, surgery, healthcare, and the issues of an aging population including recent hot topics on artificial intelligence for medicine and healthcare. The purpose of the conference is to exchange the new ideas, new technologies, and current research results in these research fields.

We received submissions from more than 10 countries. All submissions were carefully reviewed by at least two reviewers of the International Program Committee. Finally, 31 papers were accepted to be presented in this proceeding. The major areas covered at the conference and presented in this proceedings include the following: (1) Biomedical Engineering, Trends, Research, and Technologies; (2) Machine learning and labeling for biomedical visual data analysis and understanding; (3) Advanced ICT for Medical and Healthcare; (4) Healthcare Support System; and (5) Smart Medical and Healthcare System. In addition to the accepted research papers, three keynote speeches by leading researchers were presented at the conference.

We would like to thank Dr. Kyoko Hasegawa and Ms Yuka Sato of Ritsumeikan University for their valuable editing assistance for this book. We are also grateful to the authors and reviewers for their contributions.

June 2017

Yen-Wei Chen
Satoshi Tanaka
Robert J. Howlett
Lakhmi C. Jain

InMed 2017 Organization

Honorary Chair

Lakhmi C. Jain   University of Canberra, Australia
and Bournemouth University, UK

Executive Chair

Robert J. Howlett   Bournemouth University, UK

General Chair

Yen-Wei Chen   Ritsumeikan University, Japan

Program Chair

SatoshiTanaka   Ritsumeikan University, Japan

Chair of Workshop on Smart Medical and Healthcare Systems

Ivan Macia   Vicomtech-IK4, Spain

International Program Committee Members

Arnulfo Alanis	Instituto Tecnológico de Tijuana, México
Ahmad TaherAzar	Faculty of Computers and Information, Benha University, Egypt
VitoantonioBevilacqua	DEI - Politecnico di Bari, Italy
Isabelle Bichindaritz	State University of New York at Oswego, USA
Giosue' Lo Bosco	University of Palermo, Italy
Christopher Buckingham	Aston University, UK
Luis Enrique Sánchez Crespo	Grupo de Investigación GSyA, Universidad de Castilla-la Mancha, Ciudad Real, Spain
Yen-Wei Chen	Ritsumeikan University, Japan
Alex Dickinson	Faculty of Engineering and the Environment, University of Southampton, UK
Chunhua Dong	Fort Valley State University, USA
Nashwa El-Bendary	College of Management and Technology, Arab Academy for Science, Technology and Maritime Transport, Egypt
Inger Ekman	University of Gothenburg, Sweden
Massimo Esposito	Institute for High Performance Computing and Networking (ICAR) of the National Research Council of Italy (CNR), Italy
Gianluigi Ferrari	University of Parma, Italy
Kyoko Hasegawa	Ritsumeikan University, Japan
Amir Hossein Foruzan	Biomedical Engineering Group, Faculty of Engineering, Shahed University, Tehran, Iran
Luigi Gallo	National Research Council of Italy, Italy
Arnulfo Alanis Garza	Instituto Tecnológico de Tijuana, México
ArfanGhani	Coventry University, Coventry, UK
Manuel Grana	Universidad del Pais Vasco, Spain
Xianhua Han	Institute of Advance Industrial Science and Technology, Japan
Aboul Ella Hassanien	Cairo University, Egypt
Elena Hernández-Pereira	University of A Coruña, Spain
Monica Huerta	Universidad Politécnica Salesiana, Ecuador
Yutaro Iwamoto	Ritsumeikan University, Japan
Titinunt Kitrungrotsakul	Ritsumeikan University, Japan
Dalia Kriksciuniene	Vilnius University, Lithuania
Liang Li	Ritsumeikan University, Japan
Vijay Mago	Lakehead University, Canada
Bogart YailMárquezLobato	Instituto Tecnológico de Tijuana, México
José Magdaleno	Instituto Tecnológico de Tijuana, México

Esperanza Manrique	Universidad Autonoma de Baja California, México
Tadashi Matsuo	Ritsumeikan University, Japan
Rashid Mehmood	King Abdul Aziz University Jeddah, Saudi Arabia
Cheryl Metcalf	Faculty of Health Sciences, University of Southampton, UK
Nora Osuna-Millan	Universidad Autonoma del estado de Baja California, México
Hideo Miyachi	Tokyo City University, Japan
StefaniaMontani	University of Piemonte Orientale, Alessandria, Italy
Hilda Beatriz Ramírez Moreno	Universidad Autónoma de Baja California, México
AnielloMinutolo	Institute for High Performance Computing and Networking, ICAR-CNR, Italy
Stefania Montani	Università del Piemonte Orientale A. Avogadro, Italy
Kazumi Nakamatsu	School of Human Science and Environment, University of Hyogo, Japan
Haruo Noma	Ritsumeikan University, Japan
Marek Ogiela	AGH University of Science and Technology, Krakow, Poland
Andres Ortiz	Universidad de Malaga, Spain
José Sergio Magdaleno Palencia	Instituto Tecnológico de Tijuana, México
Dorin Popescu	University of Craiova, Romania
Luigi Portinale	University of Piemonte Orientale, Italy
Marco Pota	National Research Council of Italy (CNR), Institute of High Performance Computing and Networking (ICAR), Italy
Margarita RamírezRamírez	Universidad Autónoma de Baja California, México
Esperanza Manrique Rojas	Universidad Autónoma de Baja California, México
John Ronczka	SCOTTYNCC Independent Research Scientists, Australia
Yves Rybarczyk	Nova University of Lisbon (FCT/UNL), Portugal
Virgilijus Sakalauskas	Vilnius University, Lithuania
Cesar Sanin	The University of Newcastle (UoN), Australia
Catalin Stoean	University of Craiova, Romania
Ruxandra Stoean	University of Craiova, Romania
Kazuyoshi Tagawa	Ritsumeikan University, Japan
Satoshi Tanaka	Ritsumeikan University, Japan
Tomoko Tateyama	Hiroshima Institute of Technology, Intelligent Visual analytics Lab, Japan

ManolisTsiknakis	TEI Crete and FORTH, Greece
Carlos Toro	Vicomtech-IK4, Spain
Eiji Uchino	Yamaguchi University, Japan
Eloisa Vargiu	Eurecat Technology Center, Barcelona, Spain
Zhongkui Wang	Eurecat Technology Center, Barcelona, Spain
JunzoWatada	Waseda University, Japan
Wei Xiong	Institute for Infocomm Research, Singapore
Rui Xu	Dalian University of Technology, China
Yoshiyuki Yabuuchi	Shimonoseki City University, Japan

Organization and Management

KES International (www.kesinternational.org) in partnership with **the Institute of Knowledge Transfer** (www.ikt.org.uk)

Contents

Machine Learning and Labelling for Biomedical Visual Data
Analysis and Understanding

Advanced ICT for Medicine and Healthcare

Biomedical Engineering, Trends, Research and Technologies

My Health Info: An Informative mHealth App for Healthcare and Weight Control

Carlos Hurtado Sánchez[1](✉), Margarita Ramírez Ramírez[2],
José Sergio Magdaleno Palencia[1], Bogart Yail Márquez Lobato[1],
and Nora del Carmen Osuna Millán[2]

[1] Departamento de Sistemas y Computación, Instituto Tecnológico de Tijuana,
Calzada del Tecnológico S/N, Tomas Aquino, 22414 Tijuana, BC, Mexico
{carlos.hurtado,jmagdaleno,bogart}@tectijuana.edu.mx
[2] Universidad Autónoma de Baja California, Calzada Universidad 14418,
Mesa de Otay, Tijuana, Baja California, Mexico
{maguiram,nora.osuna}@uabc.edu.mx

Abstract. Health is one of the most relevant issues today, health plans and strategies for disease prevention are made in Mexico every six years, a problem is that approximately 70% of Mexicans are obese and overweight which causes diseases such as diabetes mellitus and arterial hypertension among others representing a considerable expense for health institutions in the country, to prevent this awareness, there are programs to improve population diet and physical activity. Information technologies and smartphone users are growing every day and are using mobile applications for health awareness, an informative application of health-care is developed in this paper to help with this issues, the application is divided into four sections, one is about calories contained in most common foods in the region, the next is about weight control which calculates the body mass index, ideal weight and daily water intake, other show nearby pharmacies based on your current location and last, natural treatments for hair.

Keywords: mobile Health (mHealth) · Mobile apps · Healthcare · Smart-phones

1 Introduction

Information and communication technologies (ICT) have evolved day by day and have changed the way we communicate and live, this evolution has been in large part by the use of smartphones which are the most common personal computers today, because they are always turned on, users always carry them, provide and store information in real time, are fast, have attractive interfaces and are easy to learn [1, 2].

According to the National Institute of Statistics and Geography [3], 77.7 millions of people use cell phones and two out of three users have a smartphone, being Android the dominant operating system with 62% penetration in the country [4]. In projections, nine out of every ten Mexicans will have a smartphone by 2017, due price decrease and their increment in computing power.

Another fundamental reason of smartphone adoption increment is due apps development, which has been developed for entertainment, to improve lifestyles and health among other things. One field that is constantly growing in apps development is health, as users are becoming more aware of the benefits provided to improve their lifestyle.

Studies carried out by the IMS Institute for Healthcare Informatics [5], show that there are more than 165,000 mobile applications; being fitness predominant with 36%, followed by lifestyle applications and stress management with 17% and diet and nutrition with 12%. These software tools have helped patients manage chronic degenerative diseases, their lifestyle and even self-diagnose. Approximately 80% of Android operating system applications focus on diabetic patients [6].

Health through mobile devices (mHealth), has become one of health basics aspects everywhere, including developing countries because of its portability. The doctor is no longer the only one who decides and cares for patients health, we as patients are responsible too for our longevity and life quality. Mobile apps can help us in this process in a personalized and dynamic way, providing quality information and facilitating experiences transmission, with diabetes being the therapeutic area with the greatest business potential, followed by cardiovascular diseases.

2   Common Diseases in Mexico

In Mexican Social Security Institute (IMSS) institutional program of 2014–2018 [7], the population is more exposed to a series of factors related to lifestyle such as sedentary and ingestion of industrialized foods, as well as persistence of chronic and infectious diseases. These factors have altered diseases natural history, watching transcending changes in death causes and an increment in diseases, such as diabetes mellitus, cardiovascular diseases, arterial hypertension and malignant neoplasias.

Because of this, deaths from non-transferable diseases such as diabetes mellitus, cerebrovascular diseases, hypertensive diseases and malignant neoplasias, occupied the first places in this institutional program.

The estimated cost for diseases mentioned above is 71352 million pesos approximately, this is equivalent to 30.4% of the current expenditure of health insurance and maternity, being diabetes mellitus and hypertension combined the largest proportion of these expenses with a participation of 77.9%. It is expected budget increment to treat these diseases next year.

Speaking of type 2 diabetes mellitus usually occurs in people over 40 years of age, most of whom present obesity as a pattern, concerning arterial hypertension, factors that contribute to their development are obesity, alcohol consumption, birth circumstances and work with high levels of stress. Being obesity the common denominator of both diseases.

According to [8], 70% of Mexican adults suffer obesity and one in three children; junk food and sugared drinks are the main cause of these conditions in at least half of population.

3 Types of Mobile Applications

Within mobile health, there are different apps categories of information, education, registration and monitoring, diagnostic help and treatment follow-up among others. The app presented here correspond to information category; the main objective of this kind of applications is to provide information about an illness, diet, physical activity or area of expertise to know how to treat it and how to prevent it using multimedia material.

Users should pay particular attention to this type of applications in case they recommend some medicines or food supplements since international health associations do not regulate many of these, this is why this application recommends only natural foods and shows nearby pharmacies in case that user needs a drug but does not make drug recommendations.

4 App Details

This application is about nutrition and weight control since it is one of the leading causes of diabetes, arterial hypertension and circulatory problems, which are the most common diseases in Mexico.

The application is free and in its initial phase can be downloaded for the Android operating system, to develop for iOS later, although this development is in Mexico can be used in any country of Spanish speech.

In this section, we explain technical and software aspects that were carried out for mobile development. This application was developed for the Android operating system in its early phase because according to our research most users has this operating system, Android Studio tool was used as the development environment and was programmed in Java programming language using Model View Controller (MVC) as a design pattern, a description of its components is below.

The model component is responsible for information representation needed for the application to operate, due to the fact it is informative a significant amount of time is dedicated in this phase to investigate useful information for users.

The view component is responsible for application representation or user interface; here navigation was designed together with all screens adding images, table views, and other widgets to make the interface simple, intuitive and easy to use.

The controller component takes care of application logic and establishes communication between model and view; this is where programming calculations and interaction of user interface with application information were programmed.

The developed application is fast to use and can be scalable, following the principles of software engineering and standards of good practices [9], in turn, it focuses so that anyone can use it due its interface simplicity according to design and user interaction guidelines.

5   My Health Info

My Health Info, is a health application where the user can consult information such as pharmacies, body mass index, ideal weight, hair treatment and know foods calories.

5.1   Menu

Figure 1 shows the main menu divided into four sections such as nutrition, pharmacies nearby, health and hair treatment.

Fig. 1. Main menu

5.2   Nutrition

Inside nutrition section, there is a list of foods categories that the user can consume daily, divided by alcoholic drinks, beverages, meat and derivatives and fast food shown in Fig. 2. Within each category, there is a list of foods and a description of calories contained based on its weight. Shown in Fig. 3.

Tabla de calorias
Aceites y grasas
Bebidas alcoholicas
Bebidas sin alcohol
Carnes y derivados

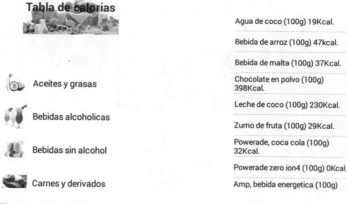

Fig. 2. Table of foods

Agua de coco (100g) 19Kcal.
Bebida de arroz (100g) 47kcal.
Bebida de malta (100g) 37Kcal.
Chocolate en polvo (100g) 398Kcal.
Leche de coco (100g) 230Kcal.
Zumo de fruta (100g) 29Kcal.
Powerade, coca cola (100g) 32Kcal.
Powerade zero ion4 (100g) 0Kcal.
Amp, bebida energetica (100g)

Fig. 3. Beverages

5.3 Pharmacies

In this section, the app will get the user location and will display it on a map, then based on its current location the nearest pharmacies will appear, as shown in Fig. 4.

Camino De La Plaza
Via Internacional
FARMACIAS SIMILARES
FARMACIAS SIMILARES
FARMACIAS SIMILARES
Av Paris
Farmacias Similares TIJUANA 31
Farmacias Similares
Tijuana
Google

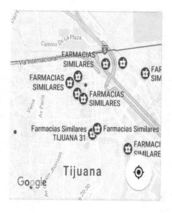

Fig. 4. Pharmacies nearby

5.4 Health

Figure 5 shows health section, this has three subsections, one to calculate body mass index BMI, another to determine the ideal weight and finally water consumption recommended according to weight.

8   C. Hurtado Sánchez et al.

Fig. 5. Health menu

BMI calculation. To calculate body mass index [10] the user must capture height in centimeters and weight in kilograms as shown in Fig. 6, then press the calculate button, and the BMI is displayed Fig. 7, Eq. (1) show the equation to calculate the BMI.

$$BMI = weight_{kg}/height_{cm}^2 \qquad (1)$$

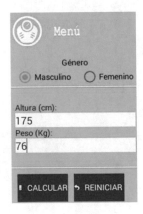

Fig. 6. BMI

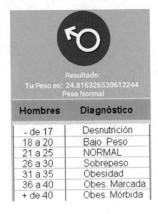

Fig. 7. BMI table

Ideal weight calculation. For the ideal weight [11] the user have to capture its gender and its height Fig. 6 and press the calculate button, Fig. 9 show the result of the calculation made. To calculate men ideal weight Eq. (2) is used and for women Eq. (3) (Fig. 6).

$$men = 50\,kg + 2.3\,kg/each\ inch\ over\ 5\,feet \tag{2}$$

$$women = 45.5\,kg + 2.3\,kg/each\ inch\ over\ 5\,feet \tag{3}$$

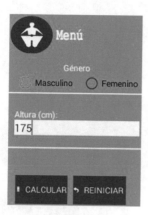

Fig. 8. Ideal weight

Fig. 9. Ideal weight result

Water consumption calculation. Figure 10 is the water consumption screen, here the user have to type its weight in kilograms to calculate [12], Fig. 11 shows users daily water intake recommended for its weight. Equation (4) shows the formula to calculate daily water intake.

$$Water\ intake = weight_{kg}/25 \tag{4}$$

Fig. 10. Water consumption

Fig. 11. Water consumption result

5.5   Hair Treatment

Figure 12 shows hair treatment section; there is special treatment for men and women.

Natural treatment. Figure 13 shows different natural treatments for men and Fig. 14 for women. As the user navigates the app, it can be treatments for:

- Hair loss.
- Hair dandruff.
- Hair growth.
- Dry hair.
- Hair shine.

Fig. 12. Hair treatment **Fig. 13.** Natural treatment   **Fig. 14.** Natural treatment for men for women

6   Conclusions

Informative mobile applications are a great help to users, thanks to them they can consult information about specific diseases and can review at any time specialized information for their treatments or conditions. In turn, it facilitates users who are not very familiar with technology since the learning curve of these apps is small and they require very few steps to observe the information they are interested in consulting, which can reach more users. It is expected that if people knows the number of calories contained in each food they can make a diet based on information presented by the application, another fundamental aspect of weight care is water consumption, it is recommended to have a weight monitor calculating their body mass index, knowing their ideal weight so they are aware of their progress and how close they are from the weight goal they want to reach. Because this application uses pure natural foods it does not endanger users health at any time.

References

1. Heron, K.E., Smyth, J.M.: Ecological momentary interventions: incorporating mobile technology into psychosocial and health behaviour treatments. Br. J. Health. Psychol. **15**(1), 1–39 (2010). doi:10.1348/135910709X466063
2. Boulos, M.N., Wheeler, S., Tavares, C., Jones, R.: How smartphones are changing the face of mobile and participatory healthcare: an overview, with example from eCAALYX. Biomed. Eng. Online **10**, 24 (2011). doi:10.1186/1475-925X-10-24
3. INEGI Instituto Nacional de Estadística y Geografia.: Estadísticas a proposito del dia mundial de internet (2016). http://www.inegi.org.mx/saladeprensa/aproposito/2016/internet2016_0.pdf
4. Observatorio Zeltia: The app intelligence: Informe 50 mejores apps de salud en español (2014). http://boletines.prisadigital.com/Informe-TAD-50-Mejores-Apps-de-Salud.pdf
5. Misra, S.: Now report finds more than 165,000 mobile health apps now available, takes close look at characteristics & use (2015). http://www.imedicalapps.com/2015/09/ims-health-apps-report/
6. Demidowich, A.P., Lu, K., Tamler, R., Bloomgarden, Z.: An evaluation of diabetes self-management applications for Android smartphones. J. Telemed. Telecare **18**(4), 235–238 (2012). doi:10.1258/jtt.2012.111002
7. IMSS Instituto Mexicano del Seguro Social: Programa institucional del Instituto Mexicano del Seguro Social (2014). http://www.imss.gob.mx/sites/all/statics/pdf/PIIMSS_2014-2018_FINAL_230414.pdf
8. Organización de las Naciones Unidas (ONU): Un 70% de los adultos en México Chile, Canadá y Estados Unidos padece obesidad (2015). http://www.un.org/spanish/News/story.asp?NewsID=31741\#.WHw8trGZOlM
9. Robertson, J., Robertson, S.: Volere. Requirements Specification Templates (2000)
10. Nuttall, F.Q.: Body Mass Index: Obesity, BMI, and Health: A Critical Review (2015). http://journals.lww.com/nutritiontodayonline/Fulltext/2015/05000/Body_Mass_Index__Obesity,_BMI,_and_Health__A.5.aspx
11. Devine, B.J.: Gentamicin therapy. Drug Intell. Clin. Pharm. **8**, 650–655 (1974)
12. Sawka, M.: Human water needs. Nutr. Rev. **63**, 30 (2005)

Big Data and Health "Clinical Records"

Margarita Ramírez Ramírez[✉], Hilda Beatriz Ramírez Moreno,
Nora del Carmen Osuna Millán, María del Consuelo Salgado Soto,
Sergio Octavio Vázquez Núñez, and Arnulfo Alanis Garza

Facultad de Contaduría y Administración, Instituto Tecnológico de Tijuana,
Universidad Autónoma de Baja California, UABC, Tijuana, BC, Mexico
{maguiram,ramirezmb,nora.osuna,csalgado,
sergio.vazquez}@uabc.edu.mx, alanis@tectijuana.edu.mx

Abstract. The health sector every day requires and implements technological tools in support prevention, prediction and treatment of diseases, as well as the capacity of the high volume of data in storage and processing of the same, that allow the health professional to give better attention and diagnose their patients. The health area integrates large amounts of data, which must be stored, classified, analyzed and consulted, all of them (with systematized and structured processes) generate useful information for the achievement of advances in health discoveries and in the administration of medical resources. BigData in medicine can have many uses and applications in areas such as epidemiology, clinical records, clinical operation, administrative management, among others.

Keywords: BigData · HealthCare · TICs

1 Introduction

The use of information and communication technologies have changed the way we do things, from an internet consultation to using technologies as work tools, are a necessity in the world in which we live. In the health area the changes that have been presented are many, to mention some from the storage of the information in medical records that allow the consultation of the information at all times until the accomplishment of surgical operations guided by technological devices. This article presents an analysis of the characteristics of technological applications in the health area, namely the use of Big Data and the implementation of clinical records. It presents the basis of BigData, its features, its functions, and existing applications, as well as the creation of a BigData model the health sector in Mexico.

2 Background

For years at a global level in the health sector, they have concentrated on curing diseases, but scientific studies have shown that most of the diseases that are currently affecting humanity are likely to be avoided, modified or controlled [1]. Hence the importance of having information of the individual from his birth and during his life, with the aim of

© Springer International Publishing AG 2018
Y.-W. Chen et al. (eds.), *Innovation in Medicine and Healthcare 2017*, Smart Innovation,
Systems and Technologies 71, DOI 10.1007/978-3-319-59397-5_2

preventing those diseases (for their food, for their way of life, for genetic issues, or for their culture etc.).

The WHO World Health Organization says in one of its principles: "The enjoyment of the highest level of health that can be achieved is one of the fundamental rights of every human being without distinction of race, religion, political ideology or economic or social condition" [2].

Having all this information in a clinical file is a difficult task and more so, in Mexico. This is because it was not until the National Development Plan of 1995–2000, which emphasized the importance of systematizing, homogenizing and updating the management of the clinical file (which contains the records of the essential technical elements for the rational study and solution of the user's health problems, involving preventive, curative and rehabilitative actions and which is constituted as a mandatory tool for the public, social and private sectors of the System National Health [3]). This Official Standard is for the Mexican national territory and its provisions are mandatory for providers of medical care services in the public, social and private sectors. Having all the information of the individual in a computer greatly benefits the doctor and the user.

2.1 TICs in Medicine

Information and communication technologies have revolutionized the way we do things without leaving out the health area. The contributions of TICs is essential in all areas but in medicine, it becomes essential because of the progress that it has. The main benefits that are obtained are better quality of care, efficiency and the reduction of costs.

The use of different devices helps in the consultation and the updating of the medical record from anywhere, having access to the information for a prompt response or making decision as the case may be, to mention an example. Other applications TICs are: clinical laboratories, hospital management software, medical records, surgeries, image files, computed tomography, nuclear magnetic resonance, medical research, to name a few.

The majority of the medical population incorporates information and communication technologies as an essential resource in their professional activities, as they offer new methods and innovations thus improving the quality of their service.

Another important factor in the health sector and ICTs are global statistics on health. The World Health Observatory (GHO) is the one that provides this information from its databases of the World Health Organization [4]. The objective is:

- Provide country data and statistics with comparable estimates
- Analysis to monitor the global, regional and national situation and trends

3 BigData

The study published by The McKinsey Global Institute (MGI) in June 2011: Define BigData as "datasets whose size goes beyond the ability to capture, store, manage and analyze database" [5].

In 2012 Gartner defined BigData as "information assets characterized by their high volume, high velocity and high variety, which demand innovative and efficient processing solutions for the improvement of knowledge and decision making in organizations." This definition makes mention of the 3 famous "V" of the BigData: Volume, Variety, and Velocity [6].

- Volume: The main benefit is the ability to process large amounts of information in large analytical data. Many companies already have large amounts of data stored, perhaps in the form of records, but do not have the ability to process them.
- Velocity: Increasing rate at which data flows into an organization has followed a similar pattern to that of volume.
- Variety: Very few times the data does itself present itself in a perfectly ordered and ready for processing. In the large information systems the data of origin are very diverse and doesn't fall into neat relational structures. It could be text from social networks, image data, a raw feed directly from a sensor source. None of these things come ready for integration into an application [7].

3.1 Project Focused on BigData

Every project that is structured in BigData focuses on the phases:

- The storage of data generated by transactional systems, external sources, integrated documents and basic information to perform information analysis.
- Processing. Technological bases to operate with large volumes of information and data flow.
- Analysis. Algorithms and methods allow to generate information from the data records and the availability of them.

3.2 Big Data and Health

BigData in medicine can have many uses and applications, in areas such as epidemiology, clinical records, clinical operation, administrative management, among others. The health area integrates large amounts of data, which must be stored, classified, analyzed and consulted, all of them with systematized and structured processes generate useful information for the achievement of advances in health discoveries and in the administration of medical resources.

In the health sector, the data on which BigData's analysis techniques can be applied are as diverse as they can be (Personal data, clinical data, administrative data). The information obtained once processed analyzed and classified the data allow the obtaining of a preventive, predictive, personalized and effective medicine.

3.3 Big Data in Preventive Medicine

Prevention is one of the relevant objectives of any health sector in any nation, prevention consists in identifying and anticipating the needs of patients, health centers or clinical laboratories, means taking steps to prevent physical, mental and sensory deficiencies (Primary prevention) or to prevent deficiencies, when they occur, have negative physical, psychological and social consequences [2].

BigData in medicine allows analyzing the data, conduct historical data studies to define specific preventive policies such as:

• Prevention health campaigns (vaccination), environmental and labor health campaigns.
• Public policies for the development of health education.
• Follow-up on epidemic outbreaks and health emergency situations.
• Development of public health research programs.

Preventive medicine is part of the data analysis to increase the social welfare of a communit.

3.4 Personalized Medicine with Big Data

BigData's techniques support the health sector by offering personalized treatment to patients by providing analytical tools and techniques for creating decision-making solutions based on real data [8].

• Encourage joint planning of patient care.
• Take advantage of information received from eSalud social networks and apps.
• Provide useful information on public health, nutritional plans, information on physical activity, etc.

The analysis of data based on BigData allows modeling the patient and provide each individual with the information and resources necessary to improve their health conditions. Therefore an effective and efficient health system.

3.5 Medical History

A tool based on the health sector, which can be identified as a set of documents generated from the relationship between the doctor and the patient, is conformed by the clinical and legal points of view of all levels of health care. The data that make up clinical history are:

• Patient filing data
• Patient medical information

According to trends, the impact of BigData on the health sector is becoming more and more important, it is becoming a growing force in the health landscape. The possibility of combining traditional data with other new forms of data both at the individual and the population level, i.e., the integration of structured and unstructured data. The health sector generates a large amount of variety of data, both structured and semi-structured. It is important to consider that structured data is data that can be stored, queried, analyzed and manipulated by computers, usually in a data table format, e.g.: The classic patient data (name, age, sex...) Structured are paper recipes, medical records, doctors 'and nurses' notes, voice recordings, X-rays, MRIs, CT scans and other medical images.

Also, electronic records of accounting and administrative management can be considered, clinical data.

The BigData, allows us to perform the transformation of data into information, support for self-care of people, support for health care providers, grouping of data.

The trend over data usage is to work with limited datasets, combine a wide variety of data, or pooling.

Medical research can make great strides with the proper management of unstructured data and organize them to define causes of illness and offer solutions (Monitoring, histories, medical treatments)

The use of BigData can be used to predict, prevent and personalize diseases, it is possible to identify its usefulness in genomic investigations, clinical administration, personalized attention to patients, patient monitoring,

3.6 Algorithmic Models for BigData in Health

Once a knowledge base of the experts and the integrated historical data have been formed, it is possible to create algorithmic models in health, that allow the treatment of information, predict the evolution of patients and their needs, and serve as Support in the practice of health professionals.

- Disease detection systems.
- Prediction of diseases and hospital readmissions.
- Early diagnosis of diseases.
- Prognosis of disease evolution and patient follow-up.
- Control of epidemics.

4   Methodology

The methodology is important for any type of process that meets the specific requirements of science [9]. A BigData is designed for the health sector which can be used as support in areas of disease prevention and detection, as well as in the follow-up of a clinical history. The phases of the methodology implemented in the construction of the BigData for health are mentioned below in Table 1 and Fig. 1.

I. Types of Users in Database
- End User: the person who uses the data, this person sees data converted into information:
- Application Developer: is the person who develops the systems that interact with the Database.
- DBA: is the person who ensures integrity, consistency, redundancy, security this is the Database Administrator who is responsible for performing daily or periodic maintenance of the data.
II. People have DBMS access are categorized as follows:
- Engineer users: they are those that interact with the system through permanent applications.

- Sophisticated users: are those with the ability to access information by means of query languages.
- Application programmers: are those with a broad domain of DML capable of generating new modules or utilities capable of handling new data in the system.
- Specialized users: are those who develop modules that do not refer to data management, but to advanced applications such as expert systems, image recognition, audio processing and so on.

Table 1. Phases in BigData design methodology in health

Storage	In this phase, the information required by the system will be integrated to identify relevant data on patients, their illnesses, treatments and physicians
Prosecution	Design of the database, determination of tables, constraints and defined data model
Analysis	This phase will allow the implementation of search algorithms and determination of patterns, for the detection of health problems

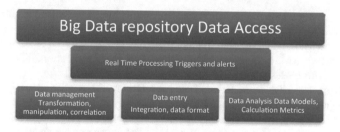

Fig. 1. Phases in BigData design methodology in health

In a BigData system, it is very important to consider the type of users involved in a system of this type, for the correct design of the system [10].

- System Users

The End User. Responsible for the health sector, and specific areas that require information and are the basis for the development of the system.

Specialized User. A user who clearly identifies the required modules and the handling of data and advanced applications such as expert systems, image recognition, audio processing, among others.

Sophisticated users. Users with the ability to access information through query tools.

- BigData design team. Team formed by experts in the design of systems, the database administrator.
- Application Developers
- BigData analytics experts.

5   Conclusions

The use of technological tools in health is becoming more useful and necessary, they offer storage and processing capabilities that allow the health professional to give better care and diagnosis to their patients. BigData implemented for the analysis and prediction of diseases are very useful, allowing advance in areas so relevant to the health sector.

These new techniques lead us to hyper personalization, which is the amount of medical data that is collected in the medical history of a person can increase exponentially and this allows us to advance in the era of knowledge where it is possible to make decisions on the basis Of the characteristics of each patient.

Likewise, thanks to the large amount of information available, it will be possible to apply artificial intelligence techniques to carry out advanced analyzes and make real-time decisions in the medical field.

References

1. Secretaria de Salud de Baja California. Tu Salud. http://www.saludbc.gob.mx/tu-salud. Accessed Jan 2017
2. Organización Mundial de la Salud. La Salud. http://www.who.int/es/. Accessed Jan 2017
3. Norma Oficial Mexicana Nom-168-Ssa1-1998, Del Expediente Clinico. http://www.salud.gob.mx/unidades/cdi/nom/168ssa18.html. Accessed Jan 2017
4. Organización Mundial de la Salud. Datos del Observatorio mundial de la salud. http://www.who.int/gho/es/pdf. Accessed Jan 2017
5. McKinsey Global Institute: Revistas-anales.es. (en línea). Accessed Sep 2016
6. Gartner: Revistas-anales.es. (en línea). http://www.revista-anales.es/web/n_29/pdf/10-16.pdf. Accessed Sep 2016
7. Dumbill, E.: Planning for Big DataA CIO's Handbook to the Changing Data Landscape, 1st edn. O'Reilly Radar Team (2012)
8. McKinsey C, Marketing: Big Data, Analytics, and the Future of Marketing & Sales, Kindle edn.
9. Bernal, C.: A: Metodología de la investigación para la administración, economía, humanidades y ciencias sociales. Pearson, Bogotá (2010)
10. Zaki, E.M.J., Wagner, M.: Data Mining and Analysis: Fundamental Concepts and Algorithms (2014)

Information and Communication Technologies, and the Positive Mental Health in the Seniors

Maricela Sevilla[1(✉)], Consuelo Salgado[1], Esperanza Manrique[1],
Hilda Beatriz Ramírez[1], and Arnulfo Alanís[2]

[1] Facultad de Contaduría y Administración, UABC, Calzada Universidad 14418,
Parque Industrial Internacional Tijuana, 22390 Tijuana, BC, Mexico
{mary_sevilla,csalgado,emanrique,ramirezmb}@uabc.edu.mx
[2] Departamento de Sistemas y Computación, Instituto Tecnológico de Tijuana,
Calzada Tecnológico S/N. Tomás Aquino, 22414 Tijuana, BC, México
alanis@tectijuana.edu.mx

Abstract. The World Health Organization indicates that the number of the elderly population is growing worldwide, with an accelerated aging of the population and an increase in life expectancy, so it is important to consider the incorporation of activities that allow Improve their physical and mental health. The objective of the article is to present the way of thinking of seniors by incorporating them into the current technological society through a process of teaching ICT, feelings, moods and what they expected before taking the course of basic computing. The results show that the participants were very motivated, they felt very confident, although with a little fear of facing the new technologies. With the passing of the weeks a change in them was noticed, greater security, and also the feeling of belonging to a group, stated that they have continued to use computers, some of them still feel insecure to carry out paperwork via the internet, but comment that They can communicate more easily with their friends and family with respect to ICT issues, and they feel part of this digital world that they once saw alien and distant.

Keywords: Active aging · Information and Communication Technologies · Mental health

1 Introduction

In traditional societies the elders had a prominent and leading role in the orientation of their respective societies. They were respected, revered, and obeyed in their role as counselors and guides to the community; they were considered repositories of the wisdom gained and accumulated throughout their life. [1], mentions in his article that this situation changed radically. In today's technologically consumed societies, particularly in underdeveloped countries, the vast majority of the elderly are victims of helplessness. From the moment they cease to be part of the productive apparatus or have an active professional life, it seems that they cease to be part of society.

Information and Communication Technologies (ICT) has become a novel tool for social interaction and mental health, so there has been a greater interest in learning how

Y.-W. Chen et al. (eds.), *Innovation in Medicine and Healthcare 2017*, Smart Innovation,
Systems and Technologies 71, DOI 10.1007/978-3-319-59397-5_3

to use them. Many of the older adults do not know how to use them or do not know the benefit that can bring them in all aspects that could be conceived in the environment. Nowadays, programs have been created focused on teaching and training them little by little about available ICTs and getting the most out of it [2].

This paper aims to show what older adults think about new technologies, moods, expectations and feelings before starting the course and what they could achieve at the end of the teaching-learning process.

2 Research Background

2.1 Active Aging

The expansion of life expectancy can be considered a success of public health policies and socioeconomic development, making the proportion of people over 60 years old increasing rapidly than any other age group in almost all countries.

But aging is also a challenge for society, which must be adapted to maximize the health and functional capacity of older people, as well as their social participation and safety, as indicated by [3].

According to [4], Mexican society must prepare itself to receive in the coming years, more than 2.5 million adults over 60 years and 700 thousand over 75 years. In this period, the participation of older adults of any age group will increase by about 50%, and within 20 years will double, both in absolute numbers and in their relative weight with respect to the total population of the country. In 2010, one in 10 Mexicans is over 60, in 2020 it will be one in six people, and by 2050 one in three.

[3] states that active aging is the "process of optimizing health, participation and safety opportunities, in order to improve the quality of life as people age."

The term active refers to a continuous participation in social, economic, cultural, spiritual and civic issues.

Active aging seeks to extend life expectancy in health and quality of life for all people as they age, including those who are fragile, disabled or in need of assistance.

Active aging, as mentioned by [5], implies that this process takes place within the context of others: friends, co-workers, neighbors and family members, where one of the important principles is Interdependence and intergenerational solidarity, that is, giving and receiving reciprocally between individuals, as well as between generations of old and young.

As we age, our body experiences various morphological and physiological changes in cognitive abilities (such as speed of learning and memory) that suffer deterioration due to lack of use, psychological as lack of motivation and in the social aspect, the Solitude and isolation, which are generated by changes of age and accumulated wear [6].

2.2 Mental Health

World Health Organization defines mental health as a state of well-being in which the individual is aware of his own abilities, can cope with the normal stresses of life, can work productively and fruitfully, and is able to make a contribution to his community.

On the other hand, [7] indicates that positive mental health is conceived based on the relationship between physical and mental aspects of people. Marie Jahoda developed a model in which she proposed how to approach Positive Metal Health through the following criteria, attitudes toward oneself, growth and self-actualization, integration, autonomy, perception of reality and mastery of the environment.

2.3 Information and Communication Technologies in the Elderly

Examining the promises of new technologies seems a prudent way of getting into the information society. Information and Communication Technologies, also known as ICT, are the set of technologies developed to manage information and send it from one place to another and encompass a wide range of solutions, including technologies to store information and retrieve it afterwards, Send and receive information from one site to another, or process information to be able to calculate results and produce reports. It is currently very high, and the Internet penetration rate is increasing worldwide.

From the above, the term "Digital divide" is used to define the differences and the separation between countries with access to new technologies and countries that do not have access to them. From another point of view, the term "Digital divide" also refers to differences between different social groups when using ICT, taking into account the different levels of literacy as well as technological capacity.

In Mexico, the Institute of Statistics, Geography and Informatics [8] published in December 2014 that the population that uses the computer in the range of 55 years or more, totals 2,171,169 users, that is, 4.4% Of 449,448,510 of the rest of the population. Regarding internet use, INEGI reports that of the 47,441,244 users, only 4.4% (2,066,906) are older than 55 years of age.

The previous data show that this sector of the population is isolated from its relatives and society by not using technological resources to communicate, socialize, update or seek information and provoke emotions that increase the feeling of loneliness.

The vast majority of seniors do not know how to use information and communication technologies, nor do they know the advantages offered by them, however, as part of different mental activation programs, programs have been developed that allow adults Greater knowledge about ICT and being part of this technological world.

3   Methodical Frameworks

The research methodology in this study is applied, qualitatively, through the case study modality. According to [9], the interest is focused on the investigation of a phenomenon, population or general condition. The study does not focus on a specific case, but on a certain set of cases for the analysis of social reality, and represents the most relevant and natural form of research oriented from a qualitative perspective. This study focuses on a particular situation, which is the situation experienced by seniors in the elderly regarding the use of technologies, their experiences at the end of the computer course and how it influences the same in their daily lives.

The sources of data collection will be the testimonies of the elderly participants and assistants to the course taught at the Faculty of Accounting and Administration, UABC, Tijuana, Mexico. The interview and survey were used as data collection techniques.

The research was carried out in three stages:

I.   Interviews to get information on what older adults expect from the computer course, their feelings about technology, fears, etc.

II. At the end of the course, the adults participating in the research project were interviewed to express their opinion about how they felt about what they learned and their level of confidence with the use of ICT.

III. After two months, a final interview was made with the participants to find out how the course influenced his daily life, his state of mind, relationship with family and friends, and so on.

4   Results and Discussion

The first survey was applied when the older adult enrolled in the course, the results of some of the questions are shown below:

The highest percentage of respondents aged 60–69 years was 54%, with ages ranging from 50 to 59 with 28%, 70 to 79 with 16.17% and 80 or more with 1.83%

The highest schooling with the highest percentage of respondents is secondary level with 37.75%, 28.30% with elementary education, 24.52% have university education and 9.43% with high school.

Regarding the occupation, 77.38% are practicing a trade, retired people represent 13.2%, 7.54% are in a profession and 1.88% do not answer.

An 84.91% of the respondents use a cell phone. 47% of the respondents said they had used the computer on several occasions, and 51% had not been able to use a computer (Fig. 1).

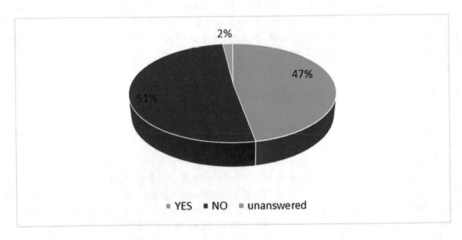

Fig. 1. Graph that shows the percentage of older adults who have used a computer.

The 86.79% of the respondents said they had internet service in their homes. As for the reason why the respondents entered this course, the outstanding answers were that 58.49% want to be updated, 47.16% to have more communication with family and friends and 33.96% to understand what people are talking about the computers (Fig. 2).

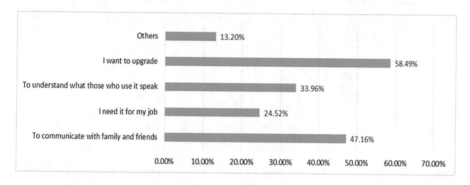

Fig. 2. Graph that indicates the reason for deciding to enter the computer course.

In the same survey there are questions to measure the mood before starting the course, some of the results were:

A 72% of respondents feel they need something in their life. Sixty percent of the respondents said they did not feel bored out of situation frequently and 38% said otherwise. 100% of those polled say they feel good spirits most of the time.

Eighty-nine percent of respondents feel energized, 75% of respondents say they do not feel worthless as they cannot do things at the same pace as before.

At the end of the course, another survey is applied, which collects information about what they think about it and about the knowledge obtained.

A 83.36% of those surveyed think that learning to use the computer has improved their self-esteem, 13.64% think that in some way learning to use computers has influenced their self-esteem, in addition, 1.5% are undecided if the course Has influenced in the improvement of their self-esteem another 1.5% did not answer.

A 89.40% of those in the group think that using the computer if it helps to update their knowledge, 6% think this contributes in a certain aspect.

The 79% think that using the computer keeps them mentally active (Fig. 3). 87.87% of respondents think that learning about technologies gives them the opportunity to develop as a person.

A 57.57% of the respondents think that with the use of the internet they feel less alone, 25.75% thinks that this helps them in certain aspect, 20.60% thinks that this helps them only a little and 4.57% that the use of the internet does not it helps me feel less alone.

After two months of completing the course, we interviewed older adults who were taken into account in the two previous surveys to know their state of mind and satisfaction regarding the use of ICTs.

The 100% of respondents think that using computers is important. Eighty-six percent of those polled say they would be able to search for anything on Google.

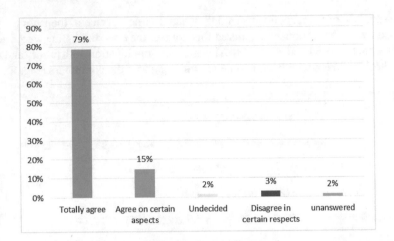

Fig. 3. Graph showing what they think about the computer holding them mentally active.

The 86% of respondents said they had not done online transactions lately, while the 14% said they had already done some paperwork online. Meanwhile, the 44% of the interviewees consider that the use of social networks seems useful, 38% consider them very useful and 18% think they are of little use.

The 92% of respondents note that using technology has helped them communicate better with their children and other family members, while 8% do not think they will help them.

The 20% think that the use of technologies has helped them to make more friends, while 80% think that they have not helped them.

5   Conclusions

With this research project it was possible to know the situation of the older adult when joining the world of ICT through a basic computer course, what he expects of him, his fears when using new technologies and the mood.

It is concluded that the majority of the participants in the course were motivated, they felt very sure of themselves, although with a little fear when facing the new technologies.

With the advancement of the course, greater security, a change of attitude towards themselves, growth and integration, and a sense of belonging to a social group with characteristics that identify older adults today were noticed.

The weekly coexistence they lived in the course allowed them to integrate into the knowledge society by learning to use ICT, leaving aside the feelings of loneliness and social and emotional isolation, and they were able to feel satisfied knowing that they can still learn, And somehow be physically and mentally active through the technologies.

After two months of completing the course, participants were again interviewed in the previous surveys and stated that they have continued to use the technologies in their daily activities; although some of them still feel insecure to perform certain activities as paperwork via the internet.

For older adults, ICT is no longer a mystery, it has become a novel tool for social interaction and mental health; Also say that they can communicate more easily with their children, grandchildren and new friends, with regard to ICT issues, and also think that they are mentally active and that they feel part of this digital world that they used to see from others and distant.

This leads us to reflect on older people who are isolated, depressed and falling out of groups that already handle some technologies. What do we have to do to integrate them into this technological society? As a family, are we aware that we should integrate them into the world of ICT?

References

1. Gómez Vecchio Ricardo: Tercera Edad y TICs. Una sociedad positiva. http://www.usuaria.org.ar/noticias/tercera-edad-y-tics-una-sociedad-positiva.html. Accessed Dec 2016
2. MetLife: La tecnología en los adultos mayores. http://w3.metlife.cl/personas/consejos-de-vida/adultos-mayores/la-tecnologia-en-los-adultos-mayores.html. Accessed January 2017
3. World Health Organization (WHO). http://www.who.int/es/. Accessed December 2016
4. INAPAN. http://www.inapam.gob.mx/work/models/INAPAM/Resource/Documentos_Inicio/Cultura_del_Envejecimiento.pdf. Accessed December 2016
5. Karina, G.V.: Envejecimiento activo, mejor vida en la tercera edad. http://www.saludymedicinas.com.mx/centros-de-salud/climaterio/prevencion/envejecimiento-activo.html. Accessed January 2017
6. Andrea, B.R., Marciel, C.P.: El envejecimiento. http://medicina.uach.cl/saludpublica/diplomado/contenido/trabajos/1/La%20Serena%202006/El_envejecimiento.pdf. Accessed December 2016
7. Edgardo, F.H.: Salud mental positiva. http://logoforo.com/salud-mental-positiva/. Accessed December 2016
8. INEGI: Institute of Statistics, Geography and Informatics. http://www.inegi.org.mx/default.aspx. Accessed December 2016
9. Sandín Esteban, M.P.: Investigación Cualitativa en educación. MacGraw-Hill, México (2003)

Prediction and Prevention of Addictions Through the Implementation of a Computational Social Simulator

Miguel Antonio Osuna Millán[1(✉)], Nora Osuna Millán[2], Esperanza Manrique Rojas[2], Maricela Sevilla Caro[2], Margarita Ramírez Ramírez[2], and Ricardo Rosales Cisneros[2]

[1] Congreso del Estado de Baja California XXII Legislatura,
Gobierno del estado de Baja California, Tijuana, BC, Mexico
miguel.osuna@congreso.gob.mx
[2] Facultad de Contaduría y Administración, Universidad Autónoma de Baja California,
UABC, Tijuana, BC, Mexico
{nora.osuna,emanrique,mary_sevilla,maguiram,
ricardorosales}@uabc.edu.mx

Abstract. The main objective of this work is to determine if a computer simulator can identify the interaction between risk factors and protective factors in relation to the emergence of substance use problems in a population of adolescents. This simulator will undoubtedly reduce time, while allowing statistics to be used to support public, private and social initiatives in order to make agile decisions in the area of health; by means of an accumulated knowledge base that is based on the experience related to research and support to the community affected or at risk of falling into the consumption of addictive substances. The prediction and prevention of addictions in Baja California and the population of Mexico in which it is used will be strengthened by this tool. The regulatory framework that establishes the prevention of addictions allows different strategies to strengthen and contribute to the Mexican government's work in the area of public health and addictions.

Keywords: Addictions · Social simulator · ICT

1 Introduction

1.1 Addictions

Addiction is a disorder, which by its manifestations can sometimes be used to justify postures, declarations, emphatic, sensed pronouncements. This disorder involves a complicated solution, which requires the articulation of different actions and from different aspects, beyond the medical-psychological treatment of the addicts. The factors that must be matched so that the disorder can arise as such are: personal factors such as genetics and personality, sociocultural factors such as symbolic value and social utility of consumption as well as the existence or not of healthy alternatives, average factors - environmental as the geographical position of a certain community and the availability of the substance or conduct and finally the pharmacological or addictive factors of a certain behavior or substance [1].

© Springer International Publishing AG 2018
Y.-W. Chen et al. (eds.), *Innovation in Medicine and Healthcare 2017*, Smart Innovation,
Systems and Technologies 71, DOI 10.1007/978-3-319-59397-5_4

1.1.1   The Health Law

This law regulates the right to the protection of one's health, based on Article 4. Of the Political Constitution of the United States of Mexico and in general terms specifies the structure that will attend to and which competes for health in the aspect of addictions.

Article 13, paragraph C, states that: "It is the responsibility of the Federation and the federal entities to prevent narcotic use, care for addictions and prosecution of crimes against health…", for this purpose, in Title XII in section Against Addictions, contains Chap. 1 and establishes the creation of the National Council against Addictions, established in Article 184 Bis, which says "The National Council Against Addictions is created, which will aim to promote and support the actions of the Public, social and private sectors aimed at preventing and combating public health problems caused by Addictions…" [2].

1.1.2   CONADIC

The main objective and mission of the National Commission Against Addictions is to promote and protect the health of Mexicans through the definition and conduct of national policy in the areas of research, prevention, treatment, training and development of human resources that support the Control of addictions, with the purpose of improving the individual, family and social quality of life [3].

The mission of the National Commission Against Addictions CONADIC is aligned with the National Development Plan and the National Health Program, included in the Ministry of Health.

1.1.3   Addiction Care in Baja California, Mexico

Baja California was one of the first states in Mexico, which institutionalized attention to addictions in all its facets. Through the Institute of psychiatry of the state of Baja California that has a staff of professionals in the area of Health focused exclusively on prevention, training, control and treatment of Addictions. Strategies have been implemented that have demonstrated progress in this area.

The Baja California Congress has implemented the Safe and Healthy Families program, where one of the main actions is to focus on the issue of addictions, and to do this creates a model of interdisciplinary care, with people who are specialized and sensitive to such problems, to attend to all the family that lives the attacks and the sequels of this disease.

According to the WHO, addiction is a physical and psychoemotional illness, which creates a dependency or need for a substance, activity or relationship. Its origin is multifactorial, so that biological, genetic, psychological and social factors are involved.

The program safe and healthy family is committed to a new model "SECOYT" (Awareness, awareness and treatment) created by the team that heads the Deputy and president of the health commission Dr. Miguel Antonio Osuna Millán. The program's main responsibility is to provide psychological, medical and psychiatric care to the communities and schools of Tijuana, Baja California, Mexico, through primary, secondary and tertiary prevention. Another important model is the so-called "therapeutic community" where a therapeutic context is created for people with drug problems, that is to say a residential context, where they coexist between them, and where the main objective is to achieve recovery through Abstinence, and psycho-emotional development that can provide you with the tools you need to stay sober for life.

To better explain what the SECOYT Model of Healthy and Safe Families Program wants to achieve with communities and schools, it is important to know the transteoric model of the Prochaska and Di Clemente Change. This is based on the basic premise that behavioral change is a process and that people have different levels of motivation, intention to change.

In order to achieve success in change, it is crucial to know in which stage the person is in relation to his problem, in order to design specific procedures that suit each subject, according to the Change Model (Fig. 1):

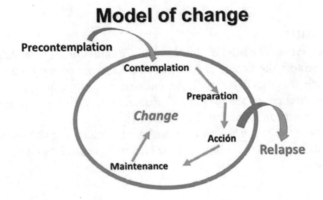

Fig. 1. Change model Prochaska and DiClemente [5–7].

Pre-contemplation: it is probable that the patient attends by external constraint (court order, relatives, etc.), expressing denial of the problem and without actually considering the change.

Contemplation: the subject recognizes having a problem, is more receptive to the information regarding their problem and possible solutions.

Preparation: at this stage the person is ready for the performance, having taken some steps towards the goal.

Action: refers to the moment when the steps taken to achieve change become more evident. Its duration is 6 months.

Maintenance: follows the action and lasts another 6 months, the purpose at this stage is to sustain the changes achieved through lifestyle modification and relapse prevention.

1.2 Behavior of the Population of Baja California, Mexico, in the Consumption of Psychoactive Substances

The State Observatory of Addictions (OAS) in the periods from 2004 to 2012, through studies conducted, observes the following pattern of behavior in the population in Baja California the prevalence of use of psychoactive substances: in 2004 there was a prevalence at the state level of 25.90%, 2005 of 33.60%, 2006 with 50.8%, in 2007 was found 47.21%, in 2008 it was obtained 36.27%, 2009 36.67% while in 2010 41.42%, decreased in 2011 to 39.44% in 2006. The highest prevalence rate was found, indicating a period of decline in consumption towards the most recent years.

These studies reflect the range of chronological age in which they began to use drugs ranging from 6 to 48 years, being between 8 and 18 years of age the largest percentage of individuals who started to use substances, being 8.2% of The respondents. The average age of start of consumption of the respondents is of 14.8 years. As for the drug with which they began to use psychoactive substances, alcohol is first with 45.8%, followed by tobacco with 17.2%, marijuana with 14.3%, sedatives, inhalants and methamphetamine with 1.5% respectively, opiates 1.0%, and ecstasy, cocaine and hallucinogens with 0.5%, respectively; The rest of the percentage mention having started consumption with combinations being 3.9% for alcohol and tobacco, 1.0% alcohol and marijuana, and 0.5% for tobacco, methamphetamine, inhalable and cocaine, 0.5% marijuana and opiates, 0.5% inhalable, marijuana and other medical drugs, marijuana and alcohol 0.5%, and 8.9% did not answer.

Due to the patterns and results observed in the studies conducted by the OAS, it is observed that there is a complex social and health problem, so with the support of information technologies, a Social Simulator will be developed to predict and prevent the use of Harmful substances and generate addictions [1, 4, 8].

1.3   Proposal to Predict and Prevent Addictions Through a Computational Social Simulator

Social simulation is adequate for the analysis of social phenomena that are inherently complex. While the idea of simulation has had enormous influence in most areas of science, and even in game programming, where there is already an emulation of societies having a significant impact on the social sciences. The progress came when they realized that computer programs offer the possibility of artificially creating societies in which people and collective actors can be directly represented organizations and observe the effect of their interactions. This provided the possibility of using experimental methods with social phenomena, and the use of computer code as a way of formalizing social dynamical theories [2].

The creation of a simulator of learning processes for the prediction and prevention of addictions, contemplates the use of technology therefore uses innovation that is a complex process, uncertain, somewhat disordered, subject to changes of many types that underlies complex dynamics And multidimensional approaches that cover a variety of actors and domains of knowledge in an environment, so that the actors in the environment must improve their skills to properly handle emergent phenomena that arise.

The creation of a process simulation tool for prediction and prevention of addictions, means that the emerging behaviors and phenomena in this domain must be modeled so that users experience in a nonlinear way, these phenomena.

The realization of this simulator will facilitate the users that will attend to the problem, through a tool that helps to experience different approaches of the emergent phenomena, in such a way that the corresponding planning and prevention for the control of addictions is developed.

1.4 Hypothesis

A computer-based social simulator can identify the interaction between risk factors and protective factors in relation to the emergence of substance use problems in a population of adolescents.

1.5 General Objective

Development of a computational social simulator for the prediction of risk factors and protective factors in relation to the appearance of substance use, problems in a population of adolescents from third year of secondary school in District X of the city of Tijuana, Baja California.

1.6 Specific Objectives

Identify prevalent risk factors of the study population.

Identify existing protective factors in the defined population.

Determine the correlation between risk factors and protective factors with substance use.

Establish the pattern of combination of risk factors and protective factors and the emergence of substance use problems.

Classify in relation to the presence of substance use in the population studied in risk groups.

Generate a computational tool that classifies the population into groups at risk.

Elaboration of the study of art on risk factors, protective factors, addiction and related elements.

Elaboration of computational modeling to determine factors associated with substance use.

Construction of a simulator that implements addiction processes.

Development of a social simulator for the prediction of addictions.

1.7 Justification

With the identification of risk factors and protective factors, the appropriate classification of groups, in high risk groups, moderate risk and low risk, may affect public health and provide the appropriate approach. The use of information and communication technologies have changed the way we do things, from an internet consultation to using technologies as work tools, are a necessity in the world in which we live. In the health area the changes that have been presented are many, to mention some from the storage of the information in medical records that allow the consultation of the information at all times until the accomplishment of surgical operations guided by technological devices. This article presents an analysis of the characteristics of technological applications in the health area, namely the use of Big Data and the implementation of clinical records. It presents the basis of BigData, its features, its functions, and existing applications, as well as the creation of a BigData model the health sector in Mexico.

2 Background and Related Work

2.1 Studies that Determine Risk Factors and Protective Factors

Research has determined how drug abuse begins and how it progresses, and there are factors that can increase a person's risk for drug abuse. Risk factors can increase a person's chances of abusing drugs while protective factors can reduce this risk. It is important to note that most people who are at risk for drug abuse do not begin to use them or become addicted, in the same way what constitutes a risk factor for one person, may not be for another.

Risk and protective factors can affect children during different stages of their lives and in each of them, risk events occur that can be changed through a preventive intervention. If untreated, negative behaviors from an early age can lead to additional risks, such as academic failure and social difficulties, which increase children's risk for future drug abuse [8].

Risk factors can influence drug abuse in a number of ways. The more risks a child is exposed to, the more likely the child is to abuse drugs. Some of the risk factors may be more powerful than others during certain stages of development, such as peer pressure during the adolescent years; As well as some protective factors, such as strong bond between parents and children, they have a greater impact on reducing risks during the early years of childhood. An important goal of prevention is to change the balance between risk factors and protection factors so that protection factors exceed those at risk [8].

3 Research Development

3.1 Computational Social Simulator

Social simulation can contribute to the understanding of social processes; Or some kind of theory or model. In general, these theories are exposed in textual form, although sometimes the theory is represented like an equation; A third way is to express theories as computer programs. Social processes can be simulated on the computer. In some circumstances, it is even possible to conduct experiments on social systems artificially that would be totally impossible or unethical to carry out in human populations [10].

Each relationship with the model must be specified exactly and each parameter; Otherwise it will be impossible to execute the simulation. This discipline also implies that the model is potentially open to inspection by other researchers, in all its details. These benefits of clarity and accuracy also have disadvantages, however. Simulations of complex social processes involve estimating many for meters and adequate data to make estimates that may be difficult to find. Another benefit of the simulation is that, in some circumstances, it can give ideas about the "appearance" of macro-level phenomena of micro-level actions. For example, a simulation of the interaction of individuals may reveal clear patterns of influence when examined on a societal scale [10].

Social simulation shows how this new methodology is adequate for the analysis of social phenomena that are inherently complex. While the idea of simulation has had enormous influence in most areas of science, and even in game programming, where there is already an emulation of societies having a significant impact on the social sciences. The breakthrough came when they realized that computer programs offer the possibility of artificially creating societies in which people and collective actors can be directly represented organizations and observe the effect of their interactions. This provided the possibility of using experimental methods with social phenomena, and the use of computer code as a way of formalizing social dynamic theories [11].

Real-world simulations including population as a target should include some means of validation. In econometrics, in political sciences and sociology data sets for verification are abundant. Other areas, mainly anthropologies suffer from a lack of data. The provision of these data are a secondary concern. The main difficulty in the datasets is appropriate to the architecture of the agent; An example of this are studies that focused mainly on the cognitive roots of social theory [12].

3.2 Methodology for Research and Development of the Tool

The methodology will be divided into two phases: The research methodology required to integrate information to form the knowledge base of the computational social simulator and the methodology required for the development of the computational social simulator.

Due to the objective and the characteristics that the research requires will be descriptive and correlational. Descriptive because it will identify psychoactive situations or facts for a diagnostic study of a social and correlational phenomenon because they will analyze the relationship or association of different variables in the study phenomenon.

The development of the research will be carried out in two stages:

(1) The first one that aims to determine the risk factors and protective factors in adolescents in the consumption of substances that are in the third year of high school in the district X (11 high schools, 16 groups with 1983 adolescents) of the city of Tijuana Baja California.
    At this stage the research will be carried out under the following phases [4, 9, 13, 14]:
    Preliminary research.
    Development conceptual framework.
    Definition of variables.
    Definition and selection of the sample.
    Analysis of the Posit data collection instrument. East
    Instrument to use allows "Self-assessment to detect problematic areas of functioning in adolescents and that could increase the risk of consumption of psychoactive substances".

Collection of data.
Analysis and interpretation of data.
Reporting of results for each period.

(2) The second stage aims to develop the computer simulator. For the development of the computer simulator, the Scrum methodology will be implemented, it is the most optimal to work collaboratively in a team, and obtain the best possible result. This methodology makes partial and regular deliveries of the final product, prioritized by the benefit they bring to the receiver of the simulator. The phases presented are as follows [15].

Start
Planning and Estimation
Implementation
Review and Retrospective
Launching

4 Conclusions

The support of a social simulator oriented to the prediction and prevention of addictions or consumption of substances harmful to health, will create an advantage in detecting and following up the risk factors and protectors that are common in various case studies, they will feed The knowledge base of the computational social simulator and will allow to create levels of artificial intelligence that generate scenarios and strategies of support in diverse cases.

The use of a digital tool, the simulator as a classifier, risk stratifier and predictor of the probability that an evaluated patient, presents substance use, in the future is certainly a social support for a community at risk.

References

1. Observatorio Estatal de las Adicciones. ipebc.gob.mx/wp-content/uploads/2016/12/oea2016.pdf. Accessed Dec 2016
2. Ley General de Salud. http://www.diputados.gob.mx/LeyesBiblio/pdf/142_161216.pdf. Accessed Nov 2016
3. Comisión Nacional Contra las Adicciones. http://www.conadic.salud.gob.mx/. Accessed Dec 2016
4. Plan anual de trabajo de la Comisión Nacional Contra las Adicciones. http://www.conadic.salud.gob.mx/pdfs/CONADIC_PAT_2016_.pdf. Accessed Nov 2016
5. DiClemente, C.C., Prochaska, J.O.: Self-change and therapy change of smoking behavior: A comparison of procecesses of change of cessation and maintenance. Addict. Behav. 7(2), 133–142 (1982)
6. Prochaska, J.: Common problems: Common solutions. Clin. Psychol. Sci. Pract. 2(1), 101–105 (1995)
7. Prochaska, J., DiClemente, C.: Transtheorical therapy: Toward a more integrative model of change. Psychother. Theory Res. Pract. 19(3), 276–288 (1982)

8. Instituto de Psiquiatría del Estado de Baja California. Observatorio Estatal Adicciones (2014). http://ipebc.gob.mx/oea2014.pdf. Accessed Nov 2016
9. National Institute of drug abuse, Advancing adiction science, Cómo prevenir el uso de drogas en los niños y los adolescentes (segunda edición). https://www.drugabuse.gov/es/pub licaciones/como-prevenir-el-uso-de-drogas/capitulo-1-los-factores-de-riesgo-y-los-factores-de-proteccion/cuales-son-los-fa. Accessed Jan 2017
10. Gilbert, N.: Computational social science: Agent-based social simulation, pp. 115– 134. Bardwell, Oxford (2007)
11. Suarez, E.D., Rodriguez-Diaz, A., Castanon-Puga, M.: Fuzzy Agents. Studies in Computational Intelligence, Vol. 154, pp. 269–293. Springer, Berlin (2008)
12. Drennan, M.: The human science of simulation: a robust hermeneutics for artificial (2005)
13. Hernandez, S.R., Fernandez, C.C., Baptista L.P.: Metodología de la Investigación 6ta. Edicion. Mc Graw Hill, México (2014)
14. Consejo Estatal contra las Adicciones de Baja California. Reglamento Interno (2004). www.ordenjuridico.gob.mx/.../Estatal/Baja%20California/wo85069.doc. Accessed January 2017
15. SCRUMstudy Tergeting sucess, SCRUM Knowledge Guide (SBOK GUIDE) (2013). http://www.scrumstudy.com/. Accessed November 2016

CFD Simulation of the Oral-Nasal Flow Partitioning During a Breathing Cycle Based on the Soft Palate Movement

Concepción Paz[1,2(✉)], Eduardo Suárez[1,2], Miguel Concheiro[1], and Marcos Conde[1]

[1] School of Industrial Engineering, University of Vigo,
Lagoas Marcosende 9, 36310 Vigo, Spain
{cpaz,suarez,mconcheiro,mfontenla}@uvigo.es
[2] Biofluids Research Group, Galicia Sur Health Research Institute (IIS Galicia Sur),
SERGAS-UVIGO, Lagoas Marcosende 9, 36310 Vigo, Spain

Abstract. The latest developments in the computational field have promoted the application of the Computational Fluid Dynamics (CFD) to the study of human health. CFD has become a new tool for in-depth investigations of the human cardiovascular and respiratory system. Therefore, this new technique provides a better understanding of the respiratory airflow, enabling the reproduction of the complex geometries of the breathing airways. In this research, a full extrathoracic model of the human airways is built from TC of an adult healthy subject. Some specific regions, as the oral cavity and the oro-pharynx, has been reconstructed from previous articles. Moreover, the soft palate has been modelled. This tissue is of main importance when considering the study of airflow patterns and the oronasal partitioning. At low ranges of physical activity, the soft palate is in anterior position allowing only nasal breathing. However, when a person begins to realize any activity, their breathing demand increases and the soft palate is displaced to a posterior position widening the oral route. The oronasal airflow partitioning was characterized on this research with five different positions of the soft palate. The breathing pattern obtained was compared with experimental data from other studies. The pressure drop and the velocity contours are analysed, to get a more detailed understanding of the breathing process.

Keywords: CFD · Oronasal partitioning · Soft palate · Extrathoracic model

1 Introduction

In the last few years, Computational Fluid Dynamics has arisen as a new tool of research in biomedical engineering. It allows the simulation of the airflow inside the respiratory airways, where experimental data is difficult to record. Moreover, it can reproduce the airflow pattern both in healthy people and in those with some respiratory disease, such as chronic obstructive pulmonary disease, asthma or obstructive sleep apnea [1–3]. In all of those cases, the breathing scheme is given by the complex geometry of the airways as well as by the characteristics of the breathing cycle: flow rate and frequency. Therefore, the flow can change from laminar to turbulent, depending on the area of study or

© Springer International Publishing AG 2018
Y.-W. Chen et al. (eds.), *Innovation in Medicine and Healthcare 2017*, Smart Innovation,
Systems and Technologies 71, DOI 10.1007/978-3-319-59397-5_5

the range of airflow simulated [4]. In a healthy subject at rest, the breathing process is done by the nasal cavity until reaching the switching point. This point is defined as the level of flow rate at which the person changes from nasal breathing to oronasal breathing. This value depends on the references consulted, and can vary between 22–44 L/min [5–7]. After this point, as the level of physical activity increases the percentage of airflow by the oral cavity begins to raise until reaching the maximum level of physical effort. However, a minimum of 40% of the total airflow should pass through the nasal airway, under any of the conditions mentioned before [8].

The breathing path and the airflow partitioning are of paramount interest when studies of particle deposition are carried out [9, 10]. The level of penetration and the concentration of delivered particles are determined by the breathing way, bearing in mind the influence of the flow rate and the airways geometry. Moreover, the effect of pollutant particles over the human health, and the delivery of drugs can be quantified with CFD.

The aim of this work is to contribute to a better understanding of this breathing process and to develop a CAD model which reproduces the airflow partitioning accuracy. To obtain a high degree of reality, the geometry is extracted from a series of CT images of an adult. Moreover, the soft palate must be added to the airways model. This tissue, that forms the posterior roof of the mouth, is the main mechanism that controls the breathing path. When a person is at rest, the soft palate is in opposition with the base of the tongue, closing the oral path. However, as the level of activity rises the soft palate moves towards the posterior wall of the pharynx, opening the oral way and narrowing the nasal one. Therefore, this movement gives advantage to oral breathing over nasal one. Different studies discuss this fact [11–13] and establish a relation between the soft palate movement and the position of the person.

2   Extrathoracic Model

According to the standard nomenclature (ICRP), the full extrathoracic airway was assembled including the nose, oral cavity, naso- and oro-pharynx, and larynx. Furthermore, the trachea and the main bronchus bifurcation were included, which was done previously [14, 15]. The used geometry is shown in detail in Fig. 1.

Our model was developed from the reconstruction of CT images of a 30 years old woman, with no respiratory disease. For the treatment of these files, in .dcm format, the software Invesalius has been used. This is a free medical program developed by the CTI (Renato Archer´s Information Technology Center), which allows the transformation of those images to .stl. This provides a faithful reconstruction of the geometry of the respiratory airways, as a cloud of points. The new files are treated in different programs until achieving 3D model in .iges format. From those images, the geometries of the nasopharynx and larynx are reconstructed using the NURBS modelling software Rhinoceros. Moreover, the glottal shape, that is the space between the vocal folds, is reproduced, in order to characterize its movement with a dynamic mesh. A wrapper process was applied in this area in order to improve the quality of elements and achieve a smoother mesh.

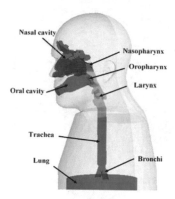

Fig. 1. 3D reconstruction of the human upper airway model.

The design of the oral cavity with the oro-pharynx is extracted from a previous article [16]. The union between this new geometry with the naso-pharinx of the TC images has been done and the palatal geometry was obtained. The palatal zone was considered as the region of intersection of both geometries. As the CT images has been taken with the subject in supine position, the tongue and the soft palate where displaced from their resting position to a posterior one. Therefore, other four positions of the soft palate were designed. In this case, the soft palate was rotated from the highest point represented as a yellow point in Fig. 2 with a negative angle for the anterior position and a positive one for the posterior positions. Moreover, its lateral walls were deformed in order to adjust them to the pharynx wall.

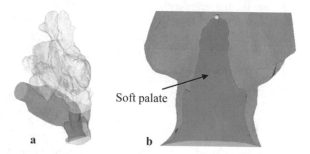

Fig. 2. Reconstruction of the soft palate, upper respiratory tract (a), and oral cavity (b).

The development of the trachea has been made as a cylinder of 12 mm of length [17, 18] with the cross section obtained from the CT images. The result of joining of all these parts is a fist model of the respiratory airways. To achieve a model with a high degree of reality the main bronchus bifurcation, and the lungs cavity were designed [19, 20]. The lungs have been modelled as a cylinder of 1600 cm^3. Moreover, the facial skin has been created and the whole model was placed inside the boundaries of a sphere of high dimensions, shown in Fig. 3. This far region will provide a constant atmospheric pressure and null velocity as it has been studied before [21].

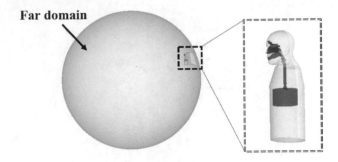

Fig. 3. Sphere that represents the far region, where the respiratory airways are located.

3   Numerical Modelling

3.1   Mesh

The mesh was generated using Hypermesh for the superficial mesh, and Fluent meshing for the volume mesh. The finite volume method used for the simulations needs the discretization of the fluid domain in computational cells. The 2D mesh of the nasal cavity was improved using the wrapping tool in Fluent. A Cartesian grid was overlaid over the original mesh. The quality was improved using the post-wrapping operation until reaching a value similar to the one of the whole corresponding mesh. A mesh of triangle elements of 1 mm was used as the superficial mesh of the respiratory airways. The size was increased gradually for the facial area and the far region. This superficial mesh was filled with 3D tetrahedral unstructured cells, apart from the lung region for which prismatic cells were used. Before establishing the element size, a convergence study was done, in order to guarantee the independence of the results from the grid size. Four different resolutions were studied, and the results were analysed in accordance with the pressure drop along the airways. In the end, a mesh of 10^6 of cells was used to conduct the simulations of this study.

3.2   Fluid Flow Modelling

Because the airflow rate considered in this study presents both a laminar and turbulent behaviour, the model selected for its characterization was the k-w SST model. This is based on the RANS approach and has integrated the Shear Stress Transport (SST), which is the most recommended model to apply with respiratory airflow [22, 23]. Different authors defend that this is the model that captures the transition between laminar and turbulent flow best. Furthermore, as the airflow rates considered here have low Mach number, the hypothesis of incompressible-air flow was applied. A pressure-based solver was configured and the pressure is related with the flow velocity throughout a correction of the pressure field. The equations were resolved sequentially, and the coupling between velocity and pressure was solved with the COUPLED scheme. Moment and turbulence were solved with the upwind scheme. The Least Squares Cell Based for resolving the gradient was used.

3.3   Boundary Conditions

In the first model, only the respiratory airways was used. In this case, a mass flow inlet was imposed at the entrance of the trachea. This condition was established with an UDF that represents the ideal form of the respiratory cycle, as a sinusoidal curve where both the inspiratory and exhalation cycle have the same length. However, conform to reality the inspiration represents between the 30–40% of the total cycle [24]. In order to get a more realistic condition, in the second model, the lungs were reproduced as a piston wall. During inspiration, the diaphragm muscles are contracted, increasing the thoracic cavity and reducing the pressure inside the lungs. The exterior air, at higher pressures, get inside the respiratory system through the nasal, oral or oronasal cavity. Thus, the lungs as a vacuum generator, allow the creation of a pressure gradient that governs the airflow, and cause the unsteadiness of the flow. This piston was configured to simulate an airflow condition between 35 L/min in the switching point up to 90 L/min at high activity. At the outlets, a pressure outlet condition was imposed both in the test model and in the last one. The main difference is that in the last model the pressure outlet condition was set in the sphere walls. This far domain reproduces a constant pressure and null velocity. This has several advantages over imposed the outlet condition at the entrance of the oral and nasal cavity [21, 25]. The boundaries that divide the airflow domain into different volumes, were considered interiors, which means that air can pass through it without any disturbances. In this group, the entrances of both oral and nasal cavities are located. The rest of the boundaries were configured as walls, with neither motion nor slip condition. No physiological consideration was taken into account, and the existence of a mucous layer was neglected. To sum up, five different cases were simulated and each one represents a different level of activity. For these cases the soft palate moved from an anterior position in the former case, to a posterior one in the latter case. However, in all of them, an oronasal airflow was considered with nasal and oral opening. The extreme cases of pure nasal or oral breathing was not the point of this article.

3.4   Convergence

For achieving the convergence of the results, a double precision simulation was done. All cases were initialized with a hybrid initialization. A first order scheme was used during the first iterations. When the solution has reached the convergence criterion, with all the residuals in a lower level than 10^{-3}, the scheme model was changed to a second order. Moreover, the flow velocity residuals was studied. The simulations were conducted with an Intel ® Xeon ® Processor E5-2697 v4 2.6 GHz cluster with 1240 GB of RAM. The model was solved using the CFD software Fluent Inc. ANSYS 17.1.

4   Results and Discussion

Five different positions of soft palate were simulated in this paper, each one corresponding to a concrete level of the respiratory airflow. The first position, the anterior one, is assigned to the flow rate in the switching point. Most people change abruptly from nasal to oronasal breathing at a flow rate around 35 L/min. This level matches a low physical activity, and depends on the studied subject. At this point, the main flow goes through the nasal cavity that guarantees a better conditioning of the air that enters the lung. However, a percentage of 43% begins to pass through the oral cavity. This percentage begins to increase until reaching a partitioning of airflow 50–50% at a flow rate of 45 L/min. After this level, the oral flow exceeds the nasal flow, at a moderate activity effort. Nevertheless, the nasal percentage stands over 40% regardless of the case studied. In this study mouth-breathers, those who breath oronasally at rest, were out of the scope.

Five different conditions of unsteady airflow, shown in Table 1, were calculated modelling the lungs as a piston. The mass flow was imposed over this membrane by applying specific movement laws. Moreover, a pressure outlet was established in the far domain. For each solution, the soft palate was positioned in a specific situation that determines the breathing partitioning. In all simulations, the oronasal breathing condition was imposed.

Table 1. Characteristics of the breathing cycle and soft palate positions simulated.

	Case 1	Case 2	Case 3	Case 4	Case 5
Q [L/min]	35	43	59.5	71.3	90
F [cycles/min]	12	18	30	31	43
Activity	*Light*	*Moderate*	*Moderate*	*High*	*Intense*

Most of previous studies assume that glottis movement has no special influence during breathing, and its movement is normally neglected. However, in this study a series of steady simulations were made in order to study the influence of the glottal motion over the respiratory airflow.

4.1 Glottal Movement

The effect of the glottis movement over the airflow partitioning was studied. It is shown that the difference between both cases is less than the 1%. Therefore, its movement is neglected in the following simulations. In Fig. 4 the oral airflow partitioning is represented for the whole respiratory cycle, both for fixed and moving glottis.

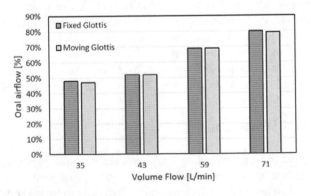

Fig. 4. Quantification of glottal effect over the airflow partitioning.

4.2   Oronasal Partitioning

The results obtained, Fig. 5, were compared with the experimental study of Niinima et al. [8]. During the two first simulations, until a volume flows around 52 L/min, the nasal pathway is the major contributor to the breathing flow. At this level, the oral and nasal curve intersects each other, agreeing with a flow partitioning of 50% oral and 50% nasal. The breathing resistance, therefore, should be the same over the two cavities. After this point, the main flow goes through the oral cavity. However, the nasal airflow stays over a higher percentage than the 39% of the total flow. This value is in accordance with previous researches. In Fig. 5 the airflow partitioning calculated as the average of the respiratory cycle is represented. The x and y axis show the volume flow in L/min and the airflow partitioning as a percentage. The difference within the experimental data is less than the 1%, which implies a high accuracy of the obtained results.

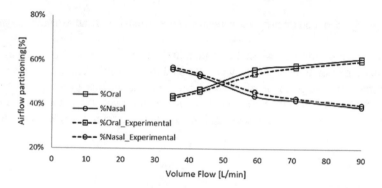

Fig. 5. Comparison between the CFD results obtained and the experimental ones [8], for the airflow partitioning.

Figures 6 (a) and (b), represent the average airflow partitioning over the two cycles. During inhalation, the oral path is the preferred one. Under this situation, the percentage of nasal flow stays under low values. However, this tendency changes in the expiratory cycle, where the nasal cavity in the desirable way until a flow rate of 59 L/min. For flows higher than this level, both oral and nasal percentages present a close value, with differences lower than the 5%. It shows that the airflow partitioning is related to the breathing cycle. Its behaviour over the exhalation is in quite an agreement with the mean of the whole cycle.

(a)   **(b)**

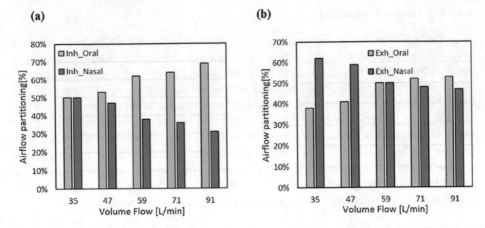

Fig. 6. Comparison of airflow partitioning between inhalation and exhalation phases.

Moreover, a relation between the angle that the geometry of soft palate was rotated, and the nasal percentage of airflow was extracted in Fig. 7 (a). A similar study was done between the percentage of nasal cross section and the airflow partitioning, Fig. 7(b). Both studies show a lineal relation between its magnitudes. To increase the airflow in the nasal cavity, a mayor cross-section, it means a lower rotated angle, is needed.

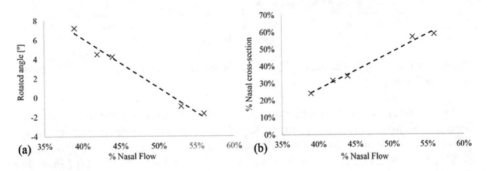

Fig. 7. Relation between the nasal flow [%] and (a) the geometric angle (b) the percentage nasal cross-section

4.3 Pressure Drop

A similar study was done in terms of total pressure in both ways, oral and nasal, calculating the pressure drop in Pa along the two cavities. Figure 8 represents the pressure drop versus the volume flow. During inhalation phase, the values of drop pressure are very similar between both cavities. However, bigger differences exist during the exhalation cycle. Under this condition, lower values were recorded for nasal exhalation, independently of flow rate. These results present a low consistency with the conclusions of the airflow partitioning during the two phases of the cycle.

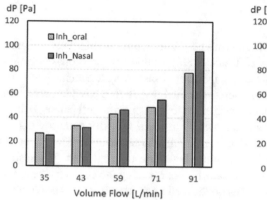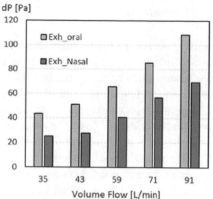

Fig. 8. Comparison of pressure drop between the inhalation and exhalation cycle.

4.4   Velocity Contours

The velocity contours of the upper respiratory airways were captured during the exhalation cycle at the same time in all the simulations. Figure 9 shows the velocity contours in the middle of sagittal plane of the respiratory airways. It shows a huge dependency between the physical effort, in terms of flow rate, and the velocity ranges. For high respiratory flows, highest velocity values were achieved in the soft palate region.

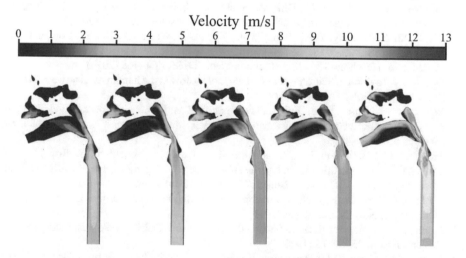

Fig. 9. Velocity contours in the middle sagittal plane during exhalation phase.

5 Conclusions

A 3D model of the full extrathoracic airways has been built from CT images of a healthy adult. Moreover, the geometry of the soft palate, responsible for the oronasal airflow partitioning, has been reproduced. Therefore, the goal of obtaining a realistic geometry of the human respiratory system to characterize the airflow patterns has been achieved. Different airflow rates were discussed. For each effort level, a specific soft palate position was studied. For low flows the palate was situated in an anterior position, narrowing the oral path and favouring the nasal flow. However, as the demands of airflow increased the soft palate was displaced to a posterior zone, widening the oral path. The results obtained were compared with experimental data, before assuring the characterization of the airflow partitioning. Notable differences on the airflow pattern were observed while comparing the average of the whole cycle, with the one of the inhalation as exhalation phase. The realistic results enable to predict the airflow patterns of individuals from CT images, and to develop a more ambitious tool for virtual surgery. For further studies, some physiological characteristics, such as the mucous layer must be included to the current model.

References

1. Luo, H., Liu, Y., Yang, X.: Particle deposition in obstructed airways. J. Biomech. **40**, 3096–3104 (2007)
2. Inthavong, K., Tu, J., Ye, Y., Ding, S., Subic, A., Thien, F.: Effects of airway obstruction induced by asthma attack on particle deposition. J. Aerosol Sci. **41**, 587–601 (2010)
3. Nie, P., Xu, X.-L., Tang, Y.-M., Wang, X.-L., Xue, X.-C., Wu, Y.-D., Zhu, M.: Computational fluid dynamics simulation of the upper airway of obstructive sleep apnea syndrome by Muller maneuver. J. Huazhong Univ. Sci. Technol. Med. Sci. **35**, 464–468 (2015)
4. Doorly, D., Taylor, D., Schroter, R.: Mechanics of airflow in the human nasal airways. Respir. Physiol. Neurobiol. **163**, 100–110 (2008)
5. Williams, J.S., Janssen, P.L., Fuller, D.D., Fregosi, R.F.: Influence of posture and breathing route on neural drive to upper airway dilator muscles during exercise. J. Appl. Physiol. **89**, 590–598 (2000)
6. Malarbet, J.L., Bertholon, J.F., Becquemin, M.H., Taieb, G., Bouchikhi, A., Roy, M.: Oral and nasal flowrate partitioning in healthy subjects performing graded exercise. Radiat. Prot. Dosimetry **53**, 179–182 (1994). Nuclear Technology Publishing
7. Lacomb, C.O.P.: Oral vs. Nasal breathing during Submaximal Aerobic Exercise. Thesis, University of Nevada, Las Vegas (2015)
8. Niinimaa, V., Cole, P., Mintz, S., Shephard, R.J.: Oronasal distribution of respiratory airflow. Respir. Physiol. **43**, 69–75 (1981)
9. Bowes, S.M., Swift, D.L.: Deposition of inhaled particles in the oral airway during oronasal breathing. Aerosol Sci. Technol. **11**, 157–167 (1989)
10. Becquemin, M.M., Bertholon, J.F., Bouchikhi, A., Malarbet, J.L., Roy, M.: Oronasal ventilation partitioning in adults and children: effect on aerosol deposition in airways. Radiat. Prot. Dosimetry. **81**, 221–228 (1999)
11. Mortimore, I.L., Marthur, R., Douglas, J.: Effect of posture, route of respiration and negative pressure on palatal muscle activity in humans. J. Appl. Physiol. **79**, 448–454 (1995)

12. Zhu, J.H., Lee, H.P., Lim, K.M., Lee, S.J., Teo, L.S.L., Wang, D.Y.: Passive movement of human soft palate during respiration: a simulation of 3D fluid/structure interaction. J. Biomech. **45**, 1992–2000 (2012)

13. Zhu, J.H, Lee, H.P., Lim, K.M., Wang, D.Y.: Simulation of movement of healthy human soft palate during respiration. In: 17th Biennial Meeting of the Canadian Society of Biomechanics/Societe Canadienne de Biomechanique (CSB/SBC), Vancouver (2012)

14. Johnstone, A., Uddin, M., Pollard, A., Heenan, A., Finlay, W.H.: The flow inside an idealised form of the human extra-thoracic airway. Exp. Fluids **37**, 673–689 (2004)

15. Malvè, M., del Palomar, A., López-Villalobos, J., Ginel, A., Doblaré, M.: FSI analysis of the coughing mechanism in a human trachea. Ann. Biomed. Eng. **38**, 1556–1565 (2010)

16. Heenan, A., Matida, E., Pollard, A., Finlay, W.: Experimental measurements and computational modeling of the flow field in an idealized human oropharynx. Exp. Fluids **35**, 70–84 (2003)

17. Seidman, P.A., Goldenberg, D., Sinz, H.S.: Tracheotomy Management. A Multidisciplinary Approach. University Cambridge Press, Cambridge (2011)

18. Cinar, U., Halezeroglu, S., Okur, E., Inanici, M.A., Kayaoglu, S.: Tracheal length in adult human: the results of 100 autopsies. Int. J. Morphol. **34**, 232–236 (2016)

19. Walters, D.K., Luke, W.H.: A method for three-dimensional navier-stokes simulations of large-scale regions of the human lung airway. J. Fluids Eng. **132**, 8 (2010)

20. Tena, A.F., Casan, P.: Use of computational fluid dynamics in respiratory medicine. Arch. Bronconeumol. **56**, 293–298 (2015)

21. Paz, C., Suárez, E., Concheiro, M., Porteiro J., Valdés, R.: CFD simulation of a CT scan oral-nasal extrathoracic model. In: Computational Methods in Multiphase Flow VII. vol. 79, pp. 387–397 (2013)

22. Kleinstreuer, C., Zhang, Z.: Laminar-to-turbulent fluid-particle flows in a human airway model. Int. J. Multiph. Flow **29**, 271–289 (2003)

23. Wen, J., Inthavong, K., Tu, J., Wang, S.: Numerical simulations for detailed airflow dynamics in a human nasal cavity. Respir. Physiol. Neurobiol. **161**, 125–135 (2008)

24. Kim, S.-H., Chung, S.-K., Na, Y.: Numerical investigation of flow-induced deformation along the human respiratory upper airway. J. Mech. Sci. Technol. **29**, 5267–5272 (2015)

25. Taylor, D.J., Doorly, D.J., Schroter, R.C.: Inflow boundary profile prescription for numerical simulation of nasal airflow. J. R. Soc. Interface **7**, 515–527 (2010)

Neural Network Backpropagation with Applications into Nutrition

A. Medina-Santiago[1(✉)], J.M. Villegas-M[2], J. Ramirez-Torres[2], N.R. García-Chong[3],
A. Cisneros-Gómez[1], E.M. Melgar-Paniagua[4], and J.I. Bermudez-Rodriguez[5]

[1] Center of Investigation, Development and Innovation Technology, University of Science
and Technology Descartes, Av. El Cipres 480, 29065 Tuxtla Gutierrez, Chiapas, Mexico
{cidit,acisneros}@universidaddescartes.edu.mx

[2] University Autonomous of Baja California, Valle of the Palms, Mexico
{villegas_josemanuel,ramirezmb}@uabc.edu.mx

[3] Medicine Faculty, University Autonomous of Chiapas, Tuxtla Gutierrez, Chiapas, Mexico
nes6g@gmail.com

[4] Institute Polytechnic National, Av. IPN, 07300 Mexico City, Mexico
emelgar@ipn.mx

[5] Institute Technology National, ITSC, Panamerican Way,
Cintalapa de Figueroa, Chiapas, Mexico
robotyks.jb@gmail.com

Abstract. This paper presents the development of a nutritional system using radial basis neural networks, that is able to provide a clear and simple prediction problems of obesity in children up to twelve years, based on your eating habits during the day. For the development of this project has taken into account various factors that are vital for the proper development of infants. A prediction system can offer a solution to several factors, which are not easily determined by conventional means. The results obtained from a sample of 186 children at primary level to obtain characteristic behaviors of the developed system are detailed in this paper. Currently, in view of the serious problem of overweight and obesity worldwide, primary schools, because of their characteristics of having a captive population and vulnerable to the benefits of education, have been identified as a suitable area for intervention studies with components to prevent this problem, considering the energy balance and the ecological models. Although there are numerous studies, at present there is no strategy that could be applied universally in schools.

Keywords: Prediction · Nutrition · Radial basis neural · Obesity · Backpropagation

1 Introduction

The ENSANUT 2006 warning about the risk in which there are more than 4 million children enter 5 to 11 years, as the combined prevalence of overweight and obesity occurs in one out of four children, also revealing that on weight and obesity has continued to increase in all ages, regions and socioeconomic groups, which has led our country to take second place in the world in obesity [1].

© Springer International Publishing AG 2018
Y.-W. Chen et al. (eds.), *Innovation in Medicine and Healthcare 2017*, Smart Innovation,
Systems and Technologies 71, DOI 10.1007/978-3-319-59397-5_6

In Mexico, based on information from two national find one built in 1999 and one in 2006, it is known that the combined prevalence of overweight and obesity has increased across the population, but particularly in the school-age population. During this period, obesity in children increased 33% and 52% children [2].

The adverse effects and risks to health obesity early in life include both short-term physical and psychosocial problems.

Longitudinal studies that childhood obesity suggest, after 3 years of age, long-term is associated with an increased risk of obesity in adulthood and with increased morbidity, persistence of associated metabolic disorders [3, 4].

School age and adolescence are a crucial configuration steps eating habits and other lifestyles that persist in later, with repercussions not only at this stage as to the possible impact as a risk factor, but also in the even in adulthood and old age.

Although there is little information regarding eating habits at school age, there are reports that report that between 7-34% of children do not eat breakfast at home before going to school [5].

Except in some private schools, public schools are not places where you can buy food or spaces where they can taste. However, playtime can buy food that sells school cooperative [6]. This name is known to the member organizations of teachers and students, staff or parents, as allowed by the regulations in force, whose role is that student's appended process of production and consumption. These cooperatives are not new and the first regulation dates from 1934, the following was issued in 1937, in 1962 appears a new, updated in 1982, which is theoretically valid [8]. It should be mentioned that the purpose of the cooperatives are essentially related to educational and learning teamwork, assimilation processes of production and consumption, the possibility that the products are sold at a lower price that prevails in the promotes the market and that earnings will be used to improve the school [7].

Nutrition is a vital issue, as vital body processes require the supply of materials and energy to provide the necessary elements for increase and repair of body tissues.

We must be aware that food is one of the pillars on which we base our health. For better or worse, food is loc is the mainstay of the formation and prevention of future disease and could be atherosclerosis, hypertension, mainly diabetic and different degrees of obesity that plague our society. Therefore, we must start from small, balanced meals, so our body will be healthy at the same time avoiding any childhood diseases [9].

There are an important aspect to consider about the power system is that this can not be generalized since each individual has different nutritional needs, that is why the alimentation depend on the activities performed by an individual throughout the day, i.e., has to be directly proportional to their activities pied otherwise fall into malnutrition [10].

In today's society, the objectives of the child alimentation have expanded and now not only aims to achieve optimal growth and prevent malnutrition and disease occurrence caracals, but also, through the same, optimizer maturational development, establish healthy habits to prevent the onset of nutritional base diseases affecting adults, trying to get a better quality of life and longevity [11–13].

As will be seen later, to secure as soon raised, we will use the radial basis neural network that has some differences from traditional logic as it defines the training and classification of data, if the theory of neural networks to predict as a data set to obtain a desired output approach, particularly the prediction data [14].

1.1   Data Collection

Problem Statement
In Mexico, the overweight and obesity is a serious public health problem affecting school children because of all socioeconomic classes.

He believes that schools primaries National System of Education, which have registered more than 95% of school-age children, are a platform that can help reverse the serious problem of overweight if it is recognized that, for now, the school is closer to being an genetic obesity environment healthy, and that part of its mission is to promote, in various educational activities, the acquisition of styles of healthy eating and physical activity. While the building has to be aimed primarily at children, also has to involve all social actors of change factors, such as parents, family, educational institutions, community authorities and, in particular, industry producing food and beverage processing, and by advertising industry print and online.

1.2   Justification

This research grew out of the need to consider one of the possible factors that cause overweight and obesity in children, and the relationship of this with the consumption of foods that are sold within the school cooperatives, hygiene, food handling, prepare hygienic measures, proper sanitation requirements and frequency of food consumption.

Hypothesis
The cooperative school meets the hygienic-Nutrient, where the food consumed within the school are a determining factor in the nutritional status of children, and applying neural network techniques can predict the behavior of infant obesity.

1.3   Theoretical Framework

This research grew out of the need to consider one of the possible factors that cause overweight and obesity in children, and the relationship of this with the consumption of foods that are sold within the school cooperatives, hygiene, food handling, prepare hygienic measures, proper sanitation requirements and frequency of food consumption.

We assessed the nutritional status of students using the criteria of the NOM-008-SSA2-1993, Control of Nutrition, Growth and Development of Children and Adolescents and evaluation form containing: Name, Grade and Group age, weight, height, Hips (cm) Waist (cm), BMI and food intake.

We evaluated the school cooperative consider through a direct observation assessment tool designed to evaluate the conditions of this, both hygienically and nutritionally about the products that are sold at recess.

Identified the relationship between the consumption of foods that is sold in the Cooperative School with nutritional status of students, as it is a very important factor for the primary students.

We developed a numerical prediction system with neural networks artificial radial basis for prediction and reinforcement of the theoretical data acquired on the job [15], obtained comparative tests with the results of observations made, based on a sheet of data and theory a biological neuron, and subsequently implement the system of artificial neural network with Backpropagation learning algorithm (Backpropagation) characterized by the Eq. (1):

$$a^4 = f^4(w^4 f^3 (w^3 f^2 (w^2 f^1 (w^1 p + b^1) + b^2) + b^3) + b^4) \tag{1}$$

1.4   Neural Networks

Neural networks consist of a simulation of the observed properties of the biological diversity of neuronal systems through mathematical models recreated through artificial mechanisms (an integrated circuit or a computer). The aim is to ensure that the machines give similar answers to which are able to give to the brain [1, 2].

A neural network consists of units called neurons, a neural network consists of units called neurons, and each neuron receives a set of inputs through interconnections and makes an exit. This output is given by three functions:

- A propagation function, which generally consists of the sum of each input multiplied by but their interconnection. If the weight is positive, the connection is called excitatory, if negative, is called inhibitor.
- An activation function that modifies the former can not exist, being here the diffusion of the same function.
- A transfer function, which applies to the value returned by the output function of the neuron and generally is given by the understanding that we can give to the outputs.

1.5   Structure

Most scientists agree that an artificial neural network ANN is very different in terms of structure of an animal brain (Fig. 1). As brain, an ANN consists of a set of simple units massively parallel processing and connections between these units where the network intelligence. Biologically the brain learns through the reorganization of synaptic connections between individual neurons [1]. Similarly, RNA having a large number of virtual processors interconnected in a simplified manner to simulate the functionality of biological neurons. In this simulation, the mechanism reorganization synaptic connections biological models with weights that are adjusted during the learning phase. An ANN trained, the weights determined set of knowledge of RNA and is capable of solving a problem.

Furthermore, each neuron is associated with a mathematical function called transfer function. This function generates the signal output from the neuron input signals. The entry of the function is the sum of all input signals but by the connection associated with that signal.

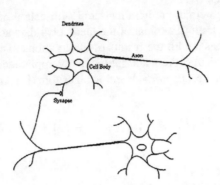

Fig. 1. Schematic drawing of biological neurons.

1.6 The Multilayer Perceptron

It basically consists of a layer of neurons with weights and adjustable threshold; this neural system may be called a neural network because existing connections in its entirety. The Perceptron training is to determine the adjustment to be performed each neuron weight to the output error is zero.

Backpropagation algorithm is a generalization of the LMS algorithm; both algorithms perform its task of updating weights and profits based on the mean square error. The network works under supervised learning and therefore requires an array of workout that will describe each output and expected output value. The structure Backpropagation neural network algorithm shown in Fig. 2.

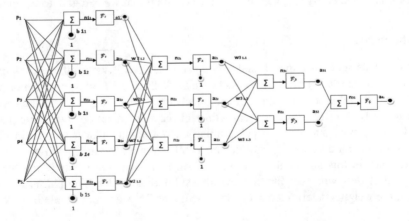

Fig. 2. Architecture multilayer perceptron with backpropagation algorithm.

1.7   Results

The Table 1 Pattern Data from the savannah of data, should be read as described promptly: results output with a neural network on equal 1.9119 and 0.9326 for girl children; regarding this outcome data savannah Girls has a degree 3 of obesity and the child would have an obesity grade 2; dependent on the age, BMI. ICC and food intake.

Table 1. Prediction result of the neural network designed and implemented in a first data with a sample of 50 children at primary school level.

Prediction of neural network backpropagation	
Girl	Boy
1.8056	0.8640
0.1258	0.8645
0.8229	0.6847
0.5027	0.8641
2.2126	0.8641
1.7740	0.8641
0.1314	0.8641
0.5692	0.8690
1.9183	0.9065
1.2402	0.8977
2.2782	0.9295
1.9119	0.9326

The following Figs. 3a to c show numerical results of the prediction system being used.

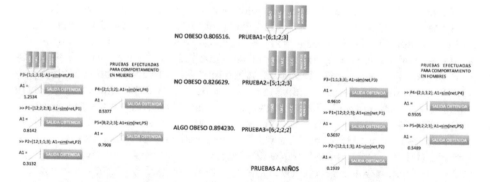

Fig. 3. a. Prediction for women with different age, BMI, WHR and food intake. 3b. Prediction for men with different age, BMI, WHR and food intake. 3c.Prediction for men with different age, BMI, WHR and food intake.

The following Figs. 4a to d shows the neuronal behavior prediction system.

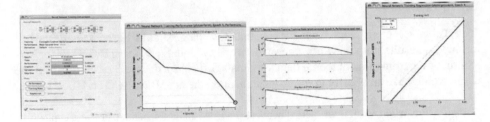

Fig. 4. a. Training of the ANN-BP. 4b. Performance of the RNA-BP in sampling error. 4c. State of training of the ANN-BP characterizing the system output. 4d.Training Regression-BP RNA.

Figure 5 shows the pseudocode for the processing of the information by Radial Base Neural Network.

```
%SISTEMA DE NUTRICION
%MTRA. LITA CAMPOS
%DR. MEDINA
%UNIVERSITY VERACRUZ
%School of Nutrition
%Institute Techonolog of Tuxtla Gutierrez
%Posgrade Biochemical and Mechatronic

%P=[EDAD DEL NIÑO,IMC,ICC,INGESTA DE ALIMENTOS]
P=[6 11;1 3;1 3;1 3];
%T=[0.70 0.85;0.78 0.94];
T=[0.70 0.85];
net1=newff(minmax(P),[5,13,22,1],{'tansig','logsig','tansig','purelin'},'traincgf');
net1.trainParam.show=5;
net1.trainParam.lr=.5;
net1.trainParam.epochs=10000;
net1.trainParam.goal=1e-3;
[net1,tr]=train(net1,P,T);
NINA=sim(net1,P)
error=T-NINA
```

Fig. 5. Pseudocode predictive system RNA-BP.

Equation characteristic of system development for prediction the nutrition is:

System Prediction = b1 + w114*tanh(conj(b1) + conj(w113)/(exp(- b1 - w112*tanh(conj(b1) + Age*conj(w11) + ingestion*conj(w14) + (mass*conj(w12))/ height + (waist*conj(w13))/hip) - w122*tanh(conj(b2) + Age*conj(w21) + ingestion*conj(w24) + (mass*conj(w22))/height + (waist*conj(w23))/hip) - w132*tanh(conj(b3) + Age*conj(w31) + ingestion*conj(w34) + (mass*conj(w32))/ height + (waist*conj(w33))/hip) - w142*tanh(conj(b4) + Age*conj(w41) + ingestion*conj(w44) + (mass*conj(w42))/height + (waist*conj(w43))/ hip)) + 1) + conj(w123)/(exp(- b2 - w212*tanh(conj(b1) + Age*conj(w11) + ingestion*conj(w14) + (mass*conj(w12))/height + (waist*conj(w13))/hip) - w222*tanh(conj(b2) + Age*conj(w21) + ingestion*conj(w24) + (mass*conj(w22))/ height + (waist*conj(w23))/hip) - w232*tanh(conj(b3) + Age*conj(w31) + ingestion*conj(w34) + (mass*conj(w32))/height + (waist*conj(w33))/hip) - w242*tanh(conj(b4) + Age*conj(w41) + ingestion*conj(w44) + (mass*conj(w42))/ height + (waist*conj(w43))/hip)) + 1) + conj(w133)/(exp(- b3 - w312*tanh(conj(b1) + Age*conj(w11) + ingestion*conj(w14) + (mass*conj(w12))/ height + (waist*conj(w13))/hip) - w322*tanh(conj(b2) + Age*conj(w21) + ingestion*conj(w24) + (mass*conj(w22))/height + (waist*conj(w23))/hip) - w332*tanh(conj(b3) + Age*conj(w31) + ingestion*conj(w34) + (mass*conj(w32))/

height + (waist*conj(w33))/hip) - w342*tanh(conj(b4) + Age*conj(w41) + ingestion*conj(w44)   +   (mass*conj(w42))/height   +   (waist*conj(w43))/
hip))   +   1))   +   w124*tanh(conj(b2)   +   conj(w213)/(exp(-   b1   -
w112*tanh(conj(b1) + Age*conj(w11) + ingestion*conj(w14) + (mass*conj(w12))/
height + (waist*conj(w13))/hip) - w122*tanh(conj(b2) + Age*conj(w21) + ingestion*conj(w24)   +   (mass*conj(w22))/height   +   (waist*conj(w23))/hip)   -
w132*tanh(conj(b3) + Age*conj(w31) + ingestion*conj(w34) + (mass*conj(w32))/
height + (waist*conj(w33))/hip) - w142*tanh(conj(b4) + Age*conj(w41) + ingestion*conj(w44)   +   (mass*conj(w42))/height   +   (waist*conj(w43))/
hip)) + 1) + conj(w223)/(exp(- b2 - w212*tanh(conj(b1) + Age*conj(w11) + ingestion*conj(w14)   +   (mass*conj(w12))/height   +   (waist*conj(w13))/hip)   -
w222*tanh(conj(b2) + Age*conj(w21) + ingestion*conj(w24) + (mass*conj(w22))/
height + (waist*conj(w23))/hip) - w232*tanh(conj(b3) + Age*conj(w31) + ingestion*conj(w34)   +   (mass*conj(w32))/height   +   (waist*conj(w33))/hip)   -
w242*tanh(conj(b4) + Age*conj(w41) + ingestion*conj(w44) + (mass*conj(w42))/
height   +   (waist*conj(w43))/hip))   +   1)   +   conj(w233)/(exp(-   b3   -
w312*tanh(conj(b1) + Age*conj(w11) + ingestion*conj(w14) + (mass*conj(w12))/
height + (waist*conj(w13))/hip) - w322*tanh(conj(b2) + Age*conj(w21) + ingestion*conj(w24)   +   (mass*conj(w22))/height   +   (waist*conj(w23))/hip)   -
w332*tanh(conj(b3) + Age*conj(w31) + ingestion*conj(w34) + (mass*conj(w32))/
height + (waist*conj(w33))/hip) - w342*tanh(conj(b4) + Age*conj(w41) + ingestion*conj(w44) + (mass*conj(w42))/height + (waist*conj(w43))/hip)) + 1))

2   Conclusions

The findings of the predictive system based on a radial basis neural network is an
effectiveness of about 99.99% of the comparative data, generating predictions on the
prospects of effective school nutrition, dependent variables mentioned above characteristics.

One explanation for the results contained in this paper is that the problem of obesity
transcends the school environment and, if the school becomes a healthy environment,
the obesogenic environments probably persist in the home and in the community where
it is possible Acquire all kinds of food and practice sedentarism as a form of well-being.
So obesity has ceased to be a personal and family matter and has become a public health
problem that requires the coordinated support of all social sectors.

References

1. Hagan, M.T., Denuth, H.B., Beale, M.: Neural Networks Design (1995)
2. Haykin, S.: Neural networks: a comprehensive foundation. Prentice Hall, Upper Saddle River
   (1999)
3. Nelso-McCord, M., Illingworth, W.T.: A Practical Guide to Neural Nets. Addison-Wesley
   Publishing Company, Inc., (Texas Instrument) (1990)
4. Principe, J.C., Euliano, N.R., Lefbvre, W.C.: Neural and Adaptive Systems Fundamentals
   Through Simulations. Wiley (1999)

5. Obesity.Implications and comprehensive management.Childhood obesity. Xenical (2013)
6. Public health n2 V47 Mexico. Cuernavaca March/April 2005. Impact escolaes a breakfast program in the prevalence of obesity and risk factors for obesity in children Sonora
7. Olaiz-Fernández, G., Rivera-Dommarco, J., Shamah-Levy, T., Rojas, R., Villalpando-Hernández, S., Herández-Avila, M., Sepúlveda-Amor, J.: National Health Survey and Nutriciñon 2006. National Public Health Institute, Cuernavaca (2006)
8. Monteiro, R., Azevedo, I.: Chronic inflammation in obesity and the metabolic syndrome. Mediators Inflamm 2010 (2010). Articulate ID 289645, doi:10.1155/2010/289645 Notas Biográficas
9. Chang, H.-T., Chen, P.-C., Huang, H.-C., Lin, D.-H.: A Study on the Application of Neural Network to the Prediction of Weight Control. Int. J. Eng. Res. Dev. 5(11), 78–85 (2013)
10. Chang, Y.-C., Lin, P.-H., Chen, C.-C., Lee, R.-G., Huang, J.-S., Tsai, T.-H.: eFurniture for home-based frailty detection using artificial neural networks and wireless sensors. Med. Eng. Phys. 35(2), 263–268 (2013)
11. Blaum, C.S., Xue, Q.L., Michelon, E., Semba, R.D., Fried, L.P.: The association between obesity and the frailty syndrome in older women: the women's health and aging studies. J. Am. Geriatr. Soc. 53, 927–934 (2005)
12. Koutkias, V.G., Chouvarda, I., Maglaveras, N.: A multiagent system enhancing home-care health services for chronic disease management. IEEE Trans. Inf. Technol. Biomed. 9, 528–537 (2005)
13. Gearhardt, A.N., Yokum, S., Stice, E., Harris, J.L., Brownell, K.D.: Relation of obesity to neural activation in response to food commercials. Soc. Cogn. Affect Neurosci. 9(7), 932–938 (2014)
14. Liang, J., Matheson, B.E., Kaye, W.H., Boutelle, K.N.: Neurocognitive correlates of obesity and obesity-related behaviors in children and adolescents. Int. J. Obes. 38, 494–506 (2014)
15. Ayatollahi, S.T.H.M.T., Zare, N.: Comparison of artificial neural networks with logistic regression for detection of obesity. J. Med. Syst. 36(4), 2449–2454 (2012)

Machine Learning and Labelling for Biomedical Visual Data Analysis and Understanding

Automatic and Robust Vessel Segmentation in CT Volumes Using Submodular Constrained Graph

Titinunt Kitrungrotsakul[1], Xian-Hua Han[1,2], Yutaro Iwamoto[1], and Yen-Wei Chen[1(✉)]

[1] Graduate School of Information Science and Engineering, Ritsumeikan University,
Kyoto, Japan
chen@is.ritsumei.ac.jp
[2] Institute of Advance Industrial Science and Technology, Tsukuba, Japan

Abstract. Graph cut is one of segmentation method that can give us the good result on natural image and large organ segmentation of medical image. However, we cannot get the correct or accurate results by using graph cut on detailed structures such as, tree branch, or blood vessel because the property of smoothness in graph cuts energy function will completely remove the small branch of the detailed structure to minimize its cost. We propose the vessel extraction method which combine graph cut and concept of submodular function. The conventional graph cuts will be use to obtain initial segmentation while graph cut with submodular function will be use to refine the initial segmentation. Submodular function can solve the problem of smoothness of graph cut in detail structure as shown in result that less segment and more united vessel tree than conventional graph cuts. The experimental result shows that our method can segment blood vessels of liver with higher accuracy while graph cut lead to a lot of loss of the detail branches in the liver vessel. With submodular constraint, we can connect the segment branch of vessel into united vessel tree which conventional graph cut still remain the segment of vessel's branches.

Keywords: Graph cuts · Submodular · Vessel segmentation · Automatic segmentation

1 Introduction

Medical image processing has been a research field attracting researcher from various field. It has been an important part and process of clinical routine. Various methods are researched and applied on computer tomography (CT), and magnetic resonance imaging (MRI), such as, segmentation, extraction, analysis, visualization, computer-aided diagnostic (CAD), and surgical planning to support doctor [1]. Blood vessel segmentation in the organ is a challenging task for medical image processing. It is really difficult to get accurate vessel segmentation results even with manually labeling by human being since blood vessels have very detailed structure. The low contrast of intensity information on the vessels and its surroundings make them hard to recognize. The segmentation of different tissues in the medical images is therefore the pre-requisite procedure for most CAD systems and visualization systems. Manually segmentation is the most

© Springer International Publishing AG 2018
Y.-W. Chen et al. (eds.), *Innovation in Medicine and Healthcare 2017*, Smart Innovation,
Systems and Technologies 71, DOI 10.1007/978-3-319-59397-5_7

accurate method in image segmentation however, it also known as the most time consuming and expensive process and extremely hard when perform on the detail structure in medical image such as vessels. Vessel segmentation methods and techniques vary depending on the image modality. Due to the variety dependency, many vessel segmentation methods are proposed.

A multi-scale approach has been widely used for vessel segmentation in various cases. Multi-scale approach uses the cylindrical structure of the vessel and segment them with a line enhancement filter with different orientation and scales. This approach is a three dimensional efficient automatic method for vessel segmentation that can be used in multiple scans and has better result with thinner slice [2]. Sato et al. [3] introduce a Three-dimensional multi-scale line filter for segmentation and visualization of curvilinear structures in medical images. The method which based on the second derivative of image using Gaussian kernel that applied on multi-scale with adaptive orientation selection using the Hessian matrix.

Graph cuts is a conventional method is now various used and developed in segmentation field on both natural and medical images [4, 5]. Graph cuts represent the image as graph $G = <\overline{V}, \overline{E}>$ where $\overline{V}$ and $\overline{E}$ correspond to nodes and edges. There are two special nodes called terminals: an "object" and "background" which used to separate the graph. A two type of user's constraints; object and background seeds; are giving to the graph by the user and those constraints are used to determine the cutting edge of the graph. Minimum cut, or min-cut, algorithm is used to divide the graph into two parts: the foreground and the background. Graph cuts are efficient finding the minimum energy and work well in not detailed structures. However, it's difficult to use those methods to segment detailed structures such as vessel.

More method and technique have been proposed for vessel extraction, such as intensity ridge traversal [6, 7], tubular filter [8, 9], skeletonization [10, 11], and active contours [12–15]. Some review can be found in [16]

The main contribution of this paper is to improve the accuracy of graph cuts method that generally cannot segment a detail structure. To improve method of graph cuts method, submodular constraint is used to improve the smoothness term of the graph cuts. The detailed of improvement method is given through the rest of this paper. The proposed graph cuts method with submodular constraint is introduced in Sect. 2. Experimental results of the proposed method and conclusions are described in Sects. 3 and 4, respectively.

2   Proposed Method

The conventional graph cut based segmentation method's energy function consisted of two properties; region and boundary property, which boundary property can consider as smoothness term. This smoothness term is the term that minimizes the cost of energy function of segmented image, which work well on many image but not in detailed structure image such as trees on natural image and vessels on medical volume. The branches of the trees and vessels are removed to minimize the cost. The required seed points for graph cuts on vessel segmentation are difficult to assign due to the structure vessels, data size, intensity information and contract. The graph cuts segmentation proposed by Jegelka [17] performed segmentation on detailed structure of natural images with high accuracy by integrated the submodular. All these graph cuts methods are generally obtained seed constraints using manual seed point from users. It is a tedious and time consuming task to use graph cuts by putting manual seed constraints for segment vessel on high-resolution medical volumes from users.

In this paper, we propose graph cut method integrates the submodular constraint to solve the problem of graph cuts smoothness term that specifically occurs on detailed structure in medical data and to improve the accuracy of segmented result. We concentrate our efforts in the following two-folds: First, the proposed method extracts the feature points from CT volume using multi-scale filter method and turning it into the seed point constraints (object and background seeds) in graph cut. As a result, it runs automatically without seed point constraints assigned by the users, which gets rid of time consuming task. Second, the proposed method integrates the submodular constraint to solve the problem of graph cuts smoothness term that specifically occurs on detailed structure in medical data and to improve the accuracy of segmented result. Figure 1, illustrates the flow diagram of the proposed framework, the flow start by using the multi-scale filter to extract background seeds and object seeds which both type of seed point will be used as constraints in graph cuts. After obtain the seed points we use graph cut to extract the initial segmentation, and use graph cuts with submodular constraint refine the segmentation result. The detail of each process will be describe in the rest of the paper.

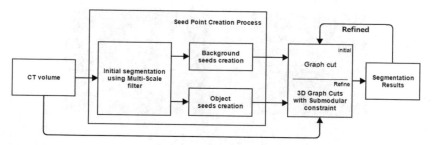

Fig. 1. The process flow diagram of the proposed automatic 3-D vessel extraction framework using graph cut and submodular constraint which consists of two main parts: (1) seed point creation process and (2) 3D graph cuts with submodular constraint.

2.1 Graph Cuts

Graph cuts arise via the relationship between cost function and set function on graph's node $G = < \overline{V}, \overline{E} >$ as aforementioned in the Sect. 1. The cost function of graph cut $E(I)$ defined in Eqs. (1) and (2) consist of two terms: (1) region term $R(A)$ and (2) boundary or smoothness term $R(B)$:

$$E(I) = \sum R(A) + \lambda \cdot \sum B(A) \tag{1}$$

$$= \sum_{p \in V} \phi_p(i_p) + \lambda \cdot \sum \phi_{p,q}(i_p, i_q) \tag{2}$$

where the coefficient λ specifies a relative importance of the region term$(B(A))$, versus smoothness term$(R(A))$ and $Pr(a|b)$ is the probability of voxel a in intensity model b. The smoothness term of graph cuts will cause problem when it is used to segment the detailed structure, which sample result of graph cut is shown in Fig. 2. Vessel region are eliminated when we add the boundary term into the cost function. Therefore, the concept of submodular will be used here to solve the problem of smoothness (min-cut Γ).

$$B_{p,q} = \exp\left(\frac{\left|i_p - i_q\right|^2}{2\sigma^2}\right) \cdot \frac{1}{d_{p,q}} \tag{3}$$

$$d_{p,q} = x_{space} \cdot \left(x_p - x_q\right)^2 + y_{space} \cdot \left(y_p - y_q\right)^2 + z_{space} \cdot \left(z_p - z_q\right)^2 \tag{4}$$

$$R_p("bkg") = \frac{\log(\Pr(i_p|"bkg"))}{\log(\Pr(i_p|"obj")) + \log(\Pr(i_p|"bkg"))} \tag{5}$$

$$R_p("obj") = \frac{\log(\Pr(i_p|"obj"))}{\log(\Pr(i_p|"obj")) + \log(\Pr(i_p|"bkg"))} \tag{6}$$

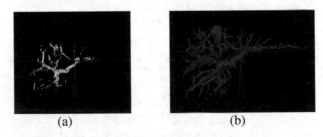

(a)   (b)

Fig. 2. (a) Result by the conventional graph cuts; (b) result of our improved graph cuts with submodular constraint.

2.2   Graph Cuts with Submodular Constraints

Submodular function is a set function whose value, informally, has the property that the difference in the incremental value of the function that a single element makes when added to an input set decreases as the size of the input set increases. In many case of image segmentation, the true boundary might have complex shape and segmentation methods such as Graph Cuts wrong identify those boundaries as background instead of object because the complexity of boundary and method try to minimize the cost function.

In the submodular function, A function $f:2^\Omega \to R$ is submodular if every $S, T \subseteq \Omega$ we have that $f(S) + f(T) \geq f(S \cap T) + f(S \cup T)$. To apply the submodular function in graph cuts, we define weighted graph $G =< \overline{V}, \overline{E}, w >$ where $\overline{V}$ and $\overline{E}$ correspond to nodes and edges of the graph cuts, respectively. $w^)$ represents the weights $w:2^E \rightarrow R$ define on subset of edge $\overline{E}$. This set function is a non-decreasing function that will used to find a new minimum cutting cost between object and background.

By integrating submodular function in to the graph cut, we can rewrite boundary term in the cost function:

$$B_{p,q} = \exp\left(\frac{\left|i_p - i_q\right|^2}{2\sigma^2}\right) \cdot \frac{1}{d_{p,q}} + \alpha \cdot f(p,q) \tag{7}$$

where

$$f(p,q) = w(\Gamma \cap S_{p,q})$$

Γ is min-cut between object and background, and $S_{p,q}$ is the class of similar edges and node. The $f(x)$ is used to reward the similarity class. To classify the class of each edge, we use the log intensity ration $log(i_p/i_q)$ and linear gradients of intensity's probability of object $R_p("obj") - R_q("obj")$ and background $R_p("bkc") - R_q("bkc")$. α is used to balance between intensity information and class edge information.

2.3 Designed Seed Points

To avoid the difficulty of assign seed point on detail structure, feature point extract is used as a tool to assign seed points in our method. The feature points are extracted from CT volume by computing vesselness probability map form. A multiscale-scale method [4] can consider as a vesselness probability map method that is used to enhance the vessel information in medical image. The normalization (vesselness) value has been proposed by Fringi [4].

$$V(\sigma) = \left\{ \begin{array}{l} 0 \ if \ \lambda_2 > 0 \ or \ \lambda_1 > 0 \\ \left(1 - \exp\left(-\frac{R_\varepsilon^2}{2\varepsilon^2}\right)\right) \exp\left(-\frac{R_\eta^2}{2\eta^2}\right)\left(1 - \exp\left(-\frac{\rho^2}{2C^2}\right)\right) \end{array} \right\} \tag{8}$$

$$V = \max_{\sigma_{min} \leq \sigma \leq \sigma_{max}} V(\sigma)$$

where ε, η, C are parameters that control the sensitivity of filter for balance between plate (R ε), blob (R η), and background (ρ), and can be formulated as follows

$$R\varepsilon = \frac{|\lambda_2|}{|\lambda_1|}, \ 0 \leq R_\eta \leq 1 \tag{9}$$

$$R_\eta = \frac{|\lambda_3|}{|\lambda_2||\lambda_1|}, \ 0 \le R\eta \le 1 \tag{10}$$

$$\rho = \sqrt{\sum_{j \le 3} \lambda_j^2}, \ 0 \le R\eta \le 1 \tag{11}$$

To select seed point vesselness image for our segmentation method, the background and object seeds can be generating by applying threshold, intensity information, intensity histogram models $Pr(A|B)$, and liver boundary with the result computed from vesselness function. Because of large intensity variance of background, we represent background intensity by using two models: (1) the darkness model consisting of air region background, liver and darkness organs and (2) the brightness model consisting of bone, spleen and brightness organs and so on. The intensity model will be used to separate the vesselness information into an initial object and background seeds. Furthermore, we refine the initial object seed by applying liver boundary and mean of intensity models as threshold (T_m). The boundary is used to limit the object seeds nearby boundary, while applying the threshold (T_b) will eliminate ambiguous seed points in both object seeds, and background seeds to obtain final seed points:

$$Seed_p = \begin{cases} obj & if & V_p > V_o, \ O_p = 1 \ and \ B_p = 0 \\ bkg & if & V_{bl} \ge V_p > V_{bh}, \ O_p = 0 \ and \ B_p = 1 \\ - & otherwise \end{cases} \tag{12}$$

$$O_p = \begin{cases} 1 & if & Pr(i_p \mid ''obj'') > T_m \\ 0 & otherwise \end{cases} \tag{13}$$

$$B_p = \begin{cases} 1 & if & Pr(i_p \mid ''bkg'') > T_m \ and \ dist(i_p, b) < T_b \\ 0 & otherwise \end{cases} \tag{14}$$

where $Pr(ip \mid ``obj'')$ and $Pr(ip \mid ``bkg'')$ are likelihood functions (intensity histograms) of the vessel (object) and liver (background). V_o is the threshold to select the object seed, the value of object seed threshold will effect to the number and quality of selected of seed points which we normally set to 0.4 to avoid the select background seed as object seed. While V_{bl} and V_{bh} are used to limit the number of background seeds which we set as 0 and 0.05, respectively.

Figure 3, show process of seed point generation of our proposed method. Figure 3(b) show the vesselness image which we apply multi-scale filter on Fig. 3(a). Figure 3(c) show the seed points after seed point generation. We can see that the object seeds (blue) are gather around center of liver which those voxels have high probability of vessel while background seeds (white) are selected around liver boundary.

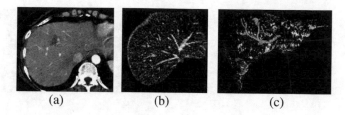

(a)   (b)   (c)

Fig. 3. (a) Original image data; (b) vesselness image after applying [2] on original image data; (c) assignment of seed points blue color is object vessel seeds and white color is background (non-vessel) seeds.

3 Experimental Results

In the experimental, we use 8 CT volumes for evaluation where all of the data have tumors as shown in Fig. 3. The manually segmented vessels are used as ground truth for quantitative evaluations. We set coefficient λ equal to 0.3, α is 0.1, T_m is 0.4, and $T_b < 5$.

Figure 4, show detailed comparisons of automatically extracted vessels and manually extracted vessels. Figure 4(a) is the manually extracted vessel, which is used as a ground truth, from the slice CT image. Figure 4(b) shows comparison of multi-scale result (red) and the ground truth (white). Figure 4(c) shows comparison of the result of conventional graph cuts (red) and the ground truth (white). Figure 4(d) shows comparison of the result of our proposed method (red) and the ground truth (white). As shown in Fig. 5, the conventional graph cuts cannot extract an accurate vascular structure, while our improved graph cuts with the submodular constraint can extract more accurate than multi-scale, and also connect more branch than conventional graph cuts.

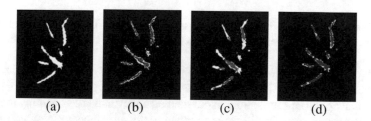

(a)   (b)   (c)   (d)

Fig. 4. The vessel segmentation results in red (a) manual segmentation in white, (b) multi-scale (vesselness) result, (c) graph cut segmentation result (d) our proposed segmentation result

Figure 5, show visualization of segmentation result from each method. Figure 5(a) is a slice CT image. Figure 5(b) is 3D visualization from multi-scale result and Fig. 5(c), and (d) are 3D vessels extracted by conventional graph cuts and our proposed method, respectively. The visualization results show that our improved graph cuts with the submodular constraint can extract a nearly perfect vascular structure.

We compare vessel extraction results generated by our proposed method and by the manual method which is used as a ground truth. The evaluation measures used in this

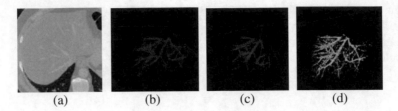

(a)   (b)   (c)   (d)

Fig. 5. The visualization of vessel segmentation result and CT data, (a) the original data, (b) the visualization multiscale result, (c) the visualization of graph cuts method's result, (d) the visualization of our proposed method's result.

experiments are the mean performance curve (Dice coefficient) as shown in Eq. (14). The experimental results are shown in Fig. 6.

$$DICE = \frac{2|A \cap B|}{|A| + |B|} \tag{15}$$

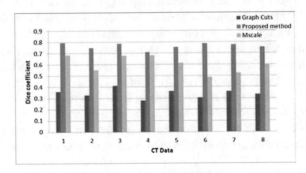

Fig. 6. Result comparison between our method and Graph Cuts: (red) result from the conventional graph cuts, (blue) our proposed method, and (green) result is multi-scale

The average accuracy of segmentation result on liver's vessels using the conventional method is only 35% while our proposed method reached 76%. It should be addressed that applying submodular in our method can enhance segmenting the detailed structure. However, there is still some problem in vessel connectivity which shown in Fig. 4(d). Comparing with conventional multi-scale filter, the best result can achieve only 0.68, lowest result is 0.48 and average result is around 0.6. Our proposed method can reach better result in all best, lowest, and average results which are 0.79, 0.71 and 0.76, respectively.

We also calculate false positive fraction (FPF) and compare our method, conventional graph cut, and multi-scale method to find liver voxel in vessel segmentation results.

$$FPF = \frac{FP}{FP + TN} \tag{16}$$

Figure 7, show that the multi-scale method has high FPF than graph cuts, and our proposed method. The comparison of our method and convention graph cut show that our method has less FPF that convention graph cut because our method used to refine the result from graph cut.

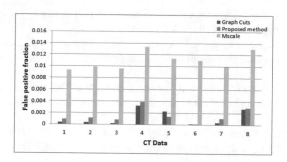

Fig. 7. False positive fraction between our method and Graph Cuts: (red) FPF from the convensional graph cuts, (blue) our proposed method, and (green) result is multi-scale

Acknowledgement. This work is supported in part by Japan Society for Promotion of Science (JSPS) under Grant No. 16J09596 and KAKEN under Grant No.15H01130, 15K00253, 16H01436.

References

1. Liang, Z.-P., Lauterbur, P.C.: Principles of Magnetic Resonance Imaging: A Signal Processing Perspective. SPIE Optical Engineering Press, New York (2000)
2. Frangi, A.F., Niessen, W.J., Vincken, K.L., Viergever, M.A.: Multiscale vessel enhancement filtering. In: MICCAI, vol. 1496, pp. 130–137 (1998)
3. Sato, Y., Nakajima, S., Atsmi, H., Koller, T., Gerig, G., Kikinis, R.: Three-dimensional multi-scale line filter for segmentation and visualization of curvilinear structures in medical images. Medical Image Analysis **2**, 143–169 (1998)
4. Boykov, Y., Funka-lea, G.: Graph cuts and efficient N-D image segmentation. Int. J. Comput. Vis. **70**, 109–131 (2006)
5. Kitrungrotsakul, T., Han, X.H., Chen, Y.W.: Liver segmentation using superpixels based graph cuts and restricted regions of shape constrains. In: IEEE International Conference on Image Processing (ICIP), pp. 3368–3371 (2015)
6. Aylward, S.R., Bullitt, E.: Initialization, noise, singularities, and scale in height ridge traversal for tubular object centerline extraction. IEEE Trans. Med. Imag. **21**, 61–75 (2002)
7. Staal, J., Abramoff, M.D., Niemeijer, M., Viergever, M.A., van Ginneken, B.: Ridge-based vessel segmentation in color images of the retina. IEEE Trans. Med. Imag. **23**, 501–508 (2004)
8. Soares, J.V.B., Leandro, J.J.G., Cesar, R.M., Jelinek, H.F., Cree, M.J.: Retinal vessel segmentation using the 2-D GaborWavelet and supervised classification. IEEE Trans. Med. Imag. **25**, 1214–1222 (2006)
9. Agam, G., Armato, S.G., Wu, C.: Vessel tree reconstruction in thoracic CT scans with application to nodule detection. IEEE Trans. Med. Imag. **24**, 486–499 (2005)
10. Yim, P.J., Choyke, P.L., Summers, R.M.: Gray-scale skeletonization of small vessels in magnetic resonance angiography. IEEE Trans. Med. Imag. **19**, 568–576 (2000)

11. Manniesing, R., Viergever, M.A., Niessen, W.J.: Vessel axis tracking using topology constrained surface evolution. IEEE Trans. Med. Imag. **26**, 309–316 (2007)
12. Lorigo, L.M., Faugeras, O.D., Grimson, W.E.L., Keriven, R., Kikinis, R., Nabavi, A., Westin, C.F.: CURVES: curve evolution for vessel segmentation. Med. Image Anal. **5**, 195–206 (2001)
13. Yan, P., Kassim, A.A.: Segmentation of volumetric MRA images by using capillary active contour. Med. Image Anal. **10**, 317–329 (2006)
14. van Bemmel, C.M., Spreeuwers, L.J., Viergever, M.A., Niessen, W.J.: Level-set-based artery-vein separation in blood pool agent CE-MR angiograms. IEEE Trans. Med. Imag. **22**, 1224–1234 (2003)
15. Manniesing, R., Velthuis, B.K., van Leeuwen, M.S., Van Der Schaaf, I.C., van Laar, P.J., Niessen, W.J.: Level set based cerebral vasculature segmentation and diameter quantification in CT angiography. Med. Image Anal. **10**, 200–214 (2006)
16. Suri, J.S., Liu, K.C., Reden, L., Laxminarayan, S.: A review on MR vascular image processing algorithms: Skeleton versus nonskeleton approaches—Part II. IEEE Trans. Inf. Tech. Biomed. **6**, 338–350 (2002)
17. Jegelka, S., Bilmes, J.: Submodularity beyond submodular energies: coupling edges in graph cuts. In: IEEE Conference on Computer Vision and Pattern Recognition (CVPR), pp. 1897–1904 (2011)

Automatic Segmentation of Cellular/Nuclear Boundaries Based on the Shape Index of Image Intensity Surfaces

Si-Hai Yang[1,2], Xian-Hua Han[3], and Yen-Wei Chen[1(✉)]

[1] School of Information Science and Engineering, Ritsumeikan University,
Kusatsu 525-8577, Japan
ysh_12@163.com, chen@is.ritsumei.ac.jp
[2] College of Computer Science and Technology, Huaqiao University, Xiamen 361021, China
[3] Faculty of Science, Yamaguchi University, Yamaguchi 753-8511, Japan
hanxh1216@gmail.com

Abstract. Segmentation of cellular and nuclear boundaries in differential interference contrast microscopy images is an important pre-processing step for biological image analysis. It is considered as a challenging task because of the interference of cell walls, blurs, nonuniform intensity background, and poor contrast between the foreground and the background. In this paper, we present a novel scheme on cellular boundary segmentation. Based on shape index (SI), the proposed method focuses on the detection of cytoplasm granules inside cellular regions. With several geometric post-processing techniques, the SI thresholding results are integrated into the segmented images. Because the size of the cytoplasm granules is usually too small comparing with the thickness of focal planes in Z-stack, we can not calculate SI values according to the method of constructing the intensity isosurface in 3D images. Consequently, we regard intensity as Z coordinate and compute SI values within each slice. A computed SI represents the shape of intensity surface or the variation of intensity near to a target pixel. Furthermore, we also show the proposed method can be applied to nuclear segmentation with a different post-processing step. Experimental results show the proposed algorithm has higher accuracy than existing schemes despite the existence of cell walls with different shapes and fluctuated intensities.

Keywords: Segmentation · Shape index · *C. elegans* · Curvature · Cellular boundary

1 Introduction

High-throughput genome-wide RNA interference (RNAi) screening is emerging as an essential tool to assist biologists in analyzing embryonic development, morphology, and phenotypes in single cell level. The large number of images produced in these studies make manual analysis intractable. Hence, automatic cellular image processing and analysis becomes an important task, where segmentation is one of the most important steps [1].

To automatically analyze the nuclear development in a 4D way based on only differential interference contrast (DIC) microscopy images, we need to trace the relative

© Springer International Publishing AG 2018
Y.-W. Chen et al. (eds.), *Innovation in Medicine and Healthcare 2017*, Smart Innovation,
Systems and Technologies 71, DOI 10.1007/978-3-319-59397-5_8

position of nuclei within cellular boundaries, which requires to compute the location changes of cells in advance during their lifetimes. To align cells among different time points, we need to segment their boundaries accurately at first. When using the RNAi screening to systematically disrupt gene expression, observing the development of *C. elegans* is a good choice. Not only because it is one of few animals which essential embryonic gene have been identified through genome-wide RNAi screening [2, 3], it also has a fixed cell lineage and a precise cell fate map [4].

However, it is difficult to achieve robust and accurate cellular boundaries by segmentation. First, microscopy images often exhibit with many blurs and potential poor contrast between the foreground and the background. Second, there exist nonuniform intensity background and significant variations on cell size and shape within cellular regions. Finally, the existence of cell walls interferes the detection of correct boundaries. Figure 1(a) and (b) show a manually segmented boundary for a cell slice, and its edge detection result using *log* operator, respectively. As we can see, there is a cell wall on the outside of the cell.

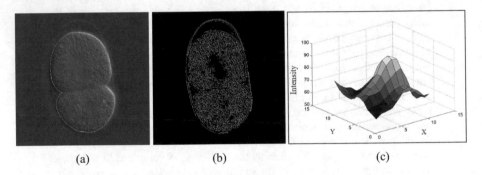

(a)   (b)   (c)

Fig. 1. A sample cell slice. (a) cell slice with manually segmented boundary (in white color), (b) the edge detection result by *log* operator, (c) the intensity surface of the small block in (a)

A large number of algorithms have been developed for automatic or semi-automatic cellular segmentation [5, 6]. These methods are based on a few basic approaches: intensity thresholding, feature detection, morphological filtering, region accumulation and deformable model fitting [1]. Learning-based schemes are also proposed in cell segmentation tasks. A notable example is the 5 categories segmentation system in cellular regions [7]. Because the learning-based methods require to label samples manually for training, which is a time-consuming and labor-intensive task. We focus on the non learning-based schemes in this study.

Methods focused on intensity thresholding are prone to be influenced by nonuniform background, although some adaptive thresholding techniques are presented [8]. Methods based on morphological filtering, especially watershed techniques, are widely used in cell segmentation [9, 10]. Nevertheless, it performs well in the segmentation of touching cells and the existence of cell walls makes it tricky to use the technique in our case. Algorithms relied on region accumulation suffers the same difficulty as the watershed method when applying it to background. Algorithms based on deformable models are also widely applied in biomedical image segmentation [11]. However, when they are used in raw images, their performances are restricted. Schemes based on feature detection are various and flexible, but they lack the overall control of the boundaries.

Kyoda *et al.* proposed an effective segmentation method based on entropy filter and the classic active contour method to segment nuclei [12] and applied the segmented nuclei to trace the development of nuclei. Although their method performs well in nuclear segmentation, its accuracy still suffers the nonuniform of background intensity when being used for cellular boundary segmentation.

Shape index (SI) is a descriptor to quantify the pure shape of a 2D surface. It was proposed by Koenderink and Doorn [13]. Its computational formula is shown in Eq. (1). Here κ_1 and κ_2 are two principal curvatures of the surface. Based on their original

$$S = \frac{2}{\pi} \arctan \frac{\kappa_2 + \kappa_1}{\kappa_2 - \kappa_1}, \; \kappa_1 \geq \kappa_2, \; S \in [-1, 1]. \tag{1}$$

work, researches developed many techniques. Cantzler *et al.* [14] exhaustively compared the properties of Gaussian curvature, mean curvature and SI. Dorai and Jain [15] established a linkage between SI and Gaussian mapping and creatively proposed shape spectrum to accumulate morphological information. Li *et al.* [16] studied a histogram on SI values and use it as a descriptor to identify facial expression. Yang *et al.* [17] showed an equivalent formula on SI and deduced the counterpart formula when there are three principal curvatures to be considered based on spherical coordinate system. Yoshida *et al.* [18] applied SI to the detection of colonic polyps in 3D images based on the intensity isosurface. Their work is adopted in many biomedical diagnostic software packages. In Fig. 1(a), as we can see, there are many cytoplasm granules in cellular regions. For each cytoplasm granule in a 2D slice, the intensity surface (Fig. 1(c)) is close to a "cap", whereas for a pixel surrounded by cytoplasm granules, the intensity surface is similar to a "cup". Therefore, shape index can be used to detect these cytoplasm granules.

In this study, a novel segmentation method for cellular boundary is presented. The algorithm focuses on the detection of massive cytoplasm granules and use shape index to detect them to avoid the interference of cell walls. Hence several geometrical postprocessing techniques are applied to obtain the cellular boundaries. Experimental results show that the proposed algorithm performs better than the existing methods. Figure 2 shows the procedure of the presented algorithm.

The remainder of this paper is organized as follow. In Sect. 2, the presented algorithm is described in details. Experimental results are presented in Sect. 3. In Sect. 4, we present an extension of the proposed method for nuclei segmentation, and Sect. 5 concludes the paper.

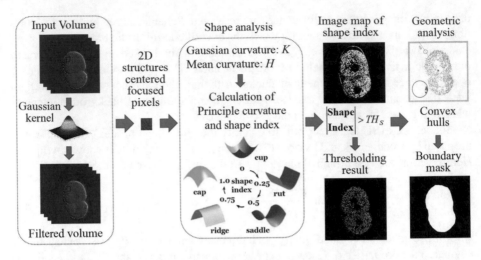

Fig. 2. The procedure of the presented algorithm

2   The Algorithm on Boundary Segmentation

In this section the proposed scheme is explained. In order to obtain the segmented cellular boundaries, the following steps are applied on each 4D DIC images:

(1) For each 2D slice $I_{t,n}$, (n: slice No. and t: time), use a 11×11 Gaussian kernel to smooth the image.
(2) For each pixel in $I_{t,n}$, compute Gaussian curvature K and mean curvature H in its 7×7 neighborhood, and then calculate the two principal curvatures κ_1, κ_2, the shape index value S, and curvedness C.
(3) Set $S = 0$ for pixels with $C < TH_c$, then threshold $I_{t,n}$ based on $|S| > TH_S$.
(4) Use geometric post-processing techniques to obtain the final segmentation results.

In the rest of this section, the above mentioned steps are described in details.

2.1   Initializing

In step (1), t is the order of time points and n is the order of z-stack. A pair of t and n defines a 2D slice. The dataset and its format are introduced on Sect. 3.1. For reducing noise, we use a Gaussian kernel to smooth each slice before computing S.

2.2   The Computation of Shape Index

In the presented algorithm, step (2) and (3) are related to the computation of SI. When detecting colonic polyps, Yoshida *et al.* used a shape index value S to represent the local shape for the intensity isosurface of a voxel based on Thirion and Gourdon's approach [19]. From this point of view, for each voxel in a 3D image, we can determine only one isosurface and calculate two principal curvatures within its neighborhood. A SI value S

thus can be calculated. This method is appropriate to detecting voxels (or pixels) located on a surface which is distinguishable from background on space and intensity. In their study, the sizes of colonic polyps to be detected range from 5–30 mm and the interval of rebuilt 3D images is 1.5–2.5 mm [18].

However, when detect cytoplasm inside cellular boundaries, using z coordinate is futile because the radii are about 0.4–0.6 μm whereas the interval of the acquired 3D images is 0.5 μm on z coordinate. As a result, it is hard to observe isosurface on z coordinate for most of cytoplasm granules.

For using SI in each slice to describe the variation of local intensity, we directly regard the intensity as z coordinate and calculate the two principal curvatures from the intensity surface for a pixel [20]. The computation of principal curvatures can use the method in [21]. Suppose $f(x,y)$ is the intensity surface over image plane (x,y), based on the first and second partial derivatives $f_x, f_y, f_{xx}, f_{yy}, f_{xy}$, we calculate Gaussian curvature with Eq. (2) and mean curvature with Eq. (3) at first, then obtain the Eqs. (5) and (6) according to Eq. (4) to compute the SI value S and the curvedness value C. In step (3), TH_S and TH_C are thresholds of S and C, respectively, here are 0.68 and 0.05.

$$K = \frac{f_{xx}f_{yy} - f_{xy}^2}{(1 + f_x^2 + f_y^2)^2},$$
(2)

$$H = \frac{f_{xx} + f_{yy} + f_{xx}f_y^2 + f_{yy}f_x^2 - 2f_xf_yf_{xy}}{(1 + f_x^2 + f_y^2)^{3/2}},$$
(3)

$$\kappa_{1,2} = H \pm \sqrt{H^2 - K}.$$
(4)

$$S = \frac{2}{\pi} \arctan\left(\frac{H}{\sqrt{H^2 - K}}\right), \quad \kappa_1 \geq \kappa_2.$$
(5)

$$C = \sqrt{\frac{1}{2}(\kappa_1^2 + \kappa_2^2)} = \sqrt{2H^2 - K}.$$
(6)

Figure 3 shows the SI detection results under this viewpoint. From the results we can see, when choosing appropriate threshold, SI can simultaneously detect the cytoplasm granules and remove most of points on cell walls.

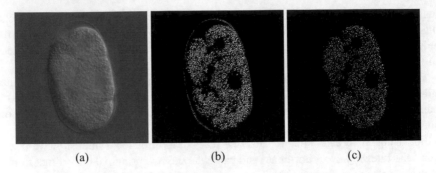

(a)	(b)	(c)

Fig. 3. A detection result. (a) a cell slice simple, (b) the result of SI thresholding with TH_S as 0.5, (c) the result of SI thresholding with TH_S as 0.7

2.3 Geometric Post-processing Techniques

After thresholding, we need to build masks from the detected points, as step (4) shows. We use three procedures to reach the goal.

First, along the direction from each boundary pixel of the 2D slice to the central of the whole detected points by SI, we use a disk model (here radius is 10) to count the number of detected points (see Fig. 4(a)). If the number is less than a threshold (here is 4), we remove these detected points. Once the number of detected pixels is larger than the threshold, we add a small piece of the disk edge into the detected points and remove other points. The goal of this step is twofold. One is to remove the noise caused by cell walls. Another is to enhance the edge points of cellular boundaries for later usage. In most cases, the added points are located inside the cellular region, thus rarely influence the final segmented boundaries.

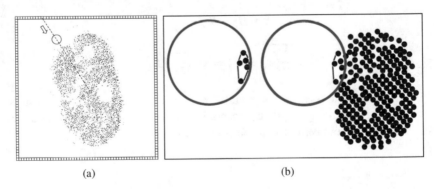

(a)	(b)

Fig. 4. A detection result. (a) the searching process, (b) the way to merge the detected points to the final boundary using convex hulls

Second, once again, along the direction from each boundary pixel P_b of the 2D slice to the central of the whole detected points, we use another disk model (here radius is 31) to built a connected cellular boundary. Once the detected points P_d within the disk is larger than a threshold (here is 20), we detect the set of points P_s located in convex hull of P_d and then remove the half part of the convex hull based on the central angle of the P_s taking P_b as the center (see Fig. 4(b)). For the remained points in P_s (at the convex hull), we bridge them in proper order.

Third, we fill the image and then detect and remove occasional protrusions. Although choosing proper threshold of SI can mainly remove the interference caused by a cell wall, there may exist few detected points at its position and occasionally not be removed by the disk-searching step, which will bring protrusions as shown in Fig. 5. To detect and mend them, for each cellular boundary point P_1 generated by the filled image, let P_2 is another boundary point along a given direction, for example clockwise, when move P_2 and make the boundary length L between P_1 and P_2 ranging from L_{min} to L_{max}, here is 5 and 100 respectively, a D/L ratio is calculated, here D is the euclidean distance between P_1 and P_2. Once the ratio is less than a threshold, here is 0.82, the piece of boundary is judged as a protrusion and a circular arc with radius 100 with the same x, y shifts is used to replace the boundary from P_1 to P_2.

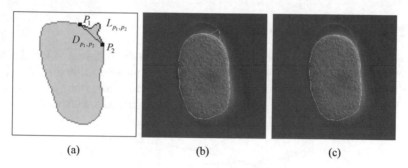

     (a)   (b)   (c)

Fig. 5. A result on the second step of post-processing. (a) a sketch map of L and D, (b) the cellular boundary after the first step, (c) the mended cellular boundary

3 Experiments

3.1 Dataset

The performance of our proposed scheme is evaluated on WDDD [12]. Using four-dimensional (4D) differential interference contrast (DIC) microscopy, WDDD has 50 sets of quantitative data from WT embryos. After silencing embryonic genes on chromosome III individually, 72 genes with 136 sets of quantitative data from RNAi experiments were also built (http://so.qbic.riken.jp/wddd). For each set, there are 180 time points (40 s/time point) and a three-dimensional (3D) image is stored at each time point. In each 3D image, the z-stack includes 66 focal planes (0.5 μm/plane) and there are 600×600 pixels (0.105 μm/pixel) in each focal plane.

3.2   Evaluation of Segmentation Performance

To evaluate the performance of the proposed scheme, we choose 41 typical slices and manually segment them as the ground truth. Then, we use entropy filter [11] and the distance regularized level set algorithm [22] (EL algorithm) as the comparison object. We also test the scheme using the adaptive thresholding method according to [8] and the morphological operators to segment cellular boundaries (TM algorithm). Besides, Watershed algorithm is also considered here. Table 1 shows the experimental results.

Table 1. The comparison of the proposed method with other algorithms

Method	Precision	Recall	Dice
EL Algorithm	0.8943	0.9005	0.8974
TM Algorithm	0.9657	0.9236	0.9442
Watershed Algorithm	0.9419	0.8400	0.8880
Proposed Algorithm	0.9737	0.9704	0.9721

3.3   Experimental Results on Different Thresholds of SI

To evaluate the performance of the proposed method under different SI thresholds, we compute the segmentation results and the precision, recall and dice ratio. when SI changing from 0.55 to 0.90 with 0.05 as the step. Figure 6 shows the result. As we can see in Fig. 6, for this study, choosing 0.65–0.7 can obtain a balanced result.

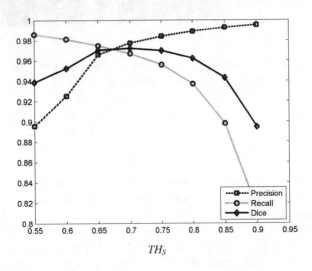

Fig. 6. The comparison result when setting different SI thresholds

4   The Application on Nuclear Segmentation

For the segmentation of nuclei using SI, because there is no interference from cell walls, we can set a low value to threshold. We use a very simple strategy to address the problem. It includes 4 steps:

(1) For each pixel in $I_{t,n}$, use $|S| < TH_S$ to threshold. (see Sect. 2 for details)
(2) In the set of detected points P, connect P_i and P_j for $\|P_i - P_j\|_2 \leq TH_d$, P_i, $P_j \in P$.
(3) Use the segmented cellular boundary from Sect. 2 to screen outside regions.
(4) Reverse the image, then apply the open operator with 6×6 disk model on each pixel in $I_{t,n}$, and remove regions with small area.

Here, $\|\cdot\|_2$ is Frobenius norm and THd is the threshold of distance. Figure 7(a) is the thresholding result on $S = 0.1$.

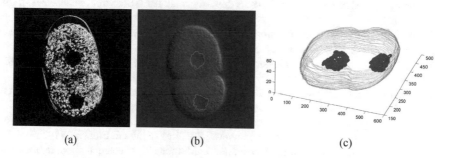

(a)   (b)   (c)

Fig. 7.   (a) The result of thresholding as $S = 0.1$. (b) The result of a slice. (c) The segmented result of a 3D image (slice No. from 15 to 55 in total 66 slices)

Figure 7(b) and (c) are the detection result for a slice and a 3D image, respectively. As we can see, using only the SI thresholding and a simple morphological operation can still obtain a good segmentation result. Han et al. using top-ranked intensity-ordered descriptors based on probability models addressed the same task [23]. The complexity of their scheme is much higher than ours.

5   Conclusion

The emergence of fully-automated phenotyping system will allow very large-scale exploratory experiments in functional genomics. Cellular boundary segmentation is a basic and indispensable step in studying morphological embryonic development. This study presented a cellular boundary segmentation algorithm based on shape index and geometric post-processing techniques. Instead of detecting cellular boundaries themselves, our method focuses on the detection of cytoplasm granules. Experimental results show high accuracy of our method. As we have shown, the proposed method can also be applied to nuclear segmentation with slight modification. As a second-order descriptor of intensity, shape index is hardly influenced by the zero-order and first-order variation of intensity, which makes it a desired tool working on images with nonuniform illumination. By contrast, thresholding techniques, most morphological operators and level set methods are based on the detection of either zero-order or first-order variation of intensity. Furthermore, because shape index is a scale-invariant descriptor, it can work on different variation range of intensity, which makes it robust when working in complex backgrounds. Experiments 3.3 shows appropriate thresholds in practical detection.

Acknowledgment. This work was supported in part by the Grant-in Aid for Scientific Research from the Japanese Ministry for Education, Science, Culture and Sports (MEXT) under the Grant No. 16H01436. The authors wish to thank Prof. Nishigawa, Dr. Onami, Dr. Tohsato and Dr. Kyoda for their advice in this research.

References

1. Meijering, E.: Cell segmentation: 50 years down the road. IEEE Sig. Proc. Magaz. **29**(5), 140–145 (2012)
2. Fraser, A.G., et al.: Functional genomic analysis of *C. elegans* chromosome I by systematic RNA interference. Nature **408**, 325–330 (2000)
3. Yang, S.H., et al.: Phenotype analysis method for identification of gene functions involved in asymmetric division of C. elegans. J. Comput. Biol. (accepted)
4. Sulston, J.E., Schierenberg, E., White, J.G., Thomson, J.N.: The embryonic cell lineage of the nematode Caenorhabditis elegans. Dev. Biol. **100**(1), 64–119 (1983)
5. Xing, F., Yang, L.: Robust nucleus/cell detection and segmentation in digital pathology and microscopy images: A comprehensive review. IEEE Rev. Biomed. Eng. **9**, 234–263 (2016)
6. Irshad, H., Veillard, A., Roux, L., Racoceanu, D.: Methods for nuclei detection, segmentation and classification in digital histopathology: A review—current status and future potential. IEEE Rev. Biomed. Eng. **7**, 97–114 (2013)
7. Ning, F., LeCun, Y., Piano, F., Bottou, L., Barbano, P.E.: Towards automatic phenotyping of developing embryos from videos. IEEE TIP **14**(9), 1360–1371 (2005)
8. Bradley, D., Roth, G.: Adaptive thresholding using the integral image. J. Graph. Tools **12**(2), 13–21 (2007)
9. Lin, G., et al.: A hybrid 3D watershed algorithm incorporating gradient cues and object models for automatic segmentation of nuclei in confocal image stacks. Cytometry A **56A**(1), 23–36 (2003)
10. Adiga, P.U., Chaudhuri, B.B.: An efficient method based on watershed and rule-based merging for segmentation of 3-D histo-pathological images. PR **34**(7), 1449–1458 (2001)
11. McInerney, T., Terzopoulos, D.: Deformable models in medical image analysis: A survey. Med. Image Anal. **1**(2), 91–108 (1996)
12. Kyoda, K., et al.: WDDD: Worm developmental dynamics database. Nucleic Acids Res. **41**(Database), 732–737 (2013)
13. Koenderink, A., Doom, V.: Surface shape and curvature scales. Image Vis. Comput. **10**(8), 557–564 (1992)
14. Cantzler, H., Fisher, R.B.: Comparison of HK and SC curvature description methods. In: Proceedings of the 3rd Conference on 3D Digital Imaging and Modeling, pp. 285–291. IEEE Computer Society Press, Los Alamitos (2001)
15. Dorai, C., Jain, A.: COSMOS - A representation scheme for 3D free-form objects. IEEE PAMI **19**(10), 1115–1130 (1997)

16. Huibin, L., Morvan, J., Liming, C.: 3D facial expression recognition based on histograms of surface differential quantities. In: Proceedings of the 13th International Conference on Advanced Concepts for Intelligent Vision Systems, pp. 483–494. Springer, Heidelberg (2011)
17. Yang, S.H., Xu, J., Suzuki, K.: Density index: Extension of the shape index in describing local intensity variations in a 3D Image. J. Comput. Aided Des. Comput. Graph. **28**(7), 1152–1159 (2016). (in Chinese)
18. Yoshida, H., Näppi, J.: Three-dimensional computer-aided diagnosis scheme for detection of colonic polyps. IEEE TMI **20**(12), 1261–1274 (2001)
19. Thirion, J.P., Gourdon, A.: Computing the differential characteristics of isointensity surfaces. CVIU **61**(2), 190–202 (1995)
20. Peet, F.G., Sahota, T.S.: Surface curvature as a measure of image texture. IEEE PAMI **7**(6), 734–738 (1985)
21. Besl, P., Jain, R.: Invariant surface characteristics for 3D object recognition in range images. CVGIP **33**(1), 33–80 (1986)
22. Li, C., Xu, C., Gui, C., Fox, M.D.: Distance regularized level set evolution and its application to image segmentation. IEEE TIP **19**(12), 3243–3254 (2010)
23. Han, X.H., Tohsato, Y., Kyoda, K., Onami, S., Nishigawa, I., Chen, Y.W.: Nuclear detection in 4D microscope images of a developing embryo using an enhanced probability map of top-ranked intensity-ordered descriptors. IPSJ Trans. Comput. Vis. Appl. **8**(8), 1–15 (2016)

A Study of Nuclei Classification Methods in Histopathological Images

Malay Singh[1,2], Zeng Zeng[3], Emarene Mationg Kalaw[2], Danilo Medina Giron[4], Kian-Tai Chong[5], and Hwee Kuan Lee[1,2,6(✉)]

[1] Department of Computer Science, School of Computing, National University of Singapore, 13 Computing Drive, Singapore 117417, Singapore
[2] Bioinformatics Institute, 30 Biopolis Street, #7-01, Matrix, Singapore 138671, Singapore
{malays,leehk}@bii.a-star.edu.sg, emakalaw@gmail.com
[3] Distributed Analytics Laboratory, Institute for Infocomm Research, 1 Fusionopolis Way, #21-01 Connexis (South Tower), Singapore 138632, Singapore
zengz@i2r.a-star.edu.sg
[4] Department of Pathology, Tan Tock Seng Hospital, 11 Jalan Tan Tock Seng, Singapore 308433, Singapore
giron_danilo@ttsh.com.sg
[5] Department of Urology, Tan Tock Seng Hospital, 11 Jalan Tan Tock Seng, Singapore 308433, Singapore
kian_tai_chong@ttsh.com.sg
[6] Image and Pervasive Access Laboratory, Institute for Infocomm Research, 1 Fusionopolis Way, #21-01 Connexis (South Tower), Singapore 138632, Singapore

Abstract. Cancer is a group of diseases involving abnormal cell growth with varying malignancy and extent across different patients. Cytological features like prominent nucleoli, nuclear enlargement, and hyperchromasia are important to the tumor pathologist in assessment of cancer malignancy from tissue biopsies. In a recent study, Yap *et al.* [21] proposed effective prominent nucleoli detectors in histopathological images and developed different feature generation methods. These methods were based on polar gradients and were used along with support vector machine (SVM) and AdaBoost for detection purposes. In this study, we benchmark the performance of these methods along with convolutional and fully connected networks for the task of distinguishing between nuclei with and without prominent nucleolus.

Keywords: AdaBoost · Auto-encoder networks · Convolutional neural networks · Deep learning · Nuclei classification

1 Introduction

Cancer is a group of diseases which involves abnormal cell growth and affects many body organs [1]. The extent and malignancy of cancer is quite different from case to case. General treatment of cancer involves examination of cancerous

© Springer International Publishing AG 2018
Y.-W. Chen et al. (eds.), *Innovation in Medicine and Healthcare 2017*, Smart Innovation, Systems and Technologies 71, DOI 10.1007/978-3-319-59397-5_9

tissues (lesions) by the pathologist, followed by treatment prescribed by the oncologist according to the pathology report.

Nucleolar changes in cancer cells are one of the cytological features important to the tumor pathologist in assessment of cancer malignancy from tissue biopsies. Abnormal nuclear morphology in cancer tissues includes nuclear enlargement, darkly stained (hyperchromatic) nuclei with irregularly clumped chromatin, and prominent nucleoli. Nuclei hence, may look different according to factors like nuclei type, nuclei life cycle, and malignancy of the disease. High grade cancer tissues generally have large nuclei with irregular color distribution and a prominent nucleolus [9]. Low grade cancer tissues generally have small, round shaped nuclei with uniform color distribution.

Prominent nucleoli patterns are one of the important features used by pathologists in their assessment. Yap et al. [21] developed a prominent nucleoli detector as an automated tool to aid the pathologist. This study proposed various feature generation methods which used nuclear feature information and compared their performance for the prominent nucleoli detection task. We can define the detection task as distinguishing nucleus with a prominent nucleolus from all the other patterns like stroma, cytoplasm, lumen, nuclei without prominent nucleolus, etc. in the histopathological images.

Along with various hand-crafted features, recent studies indicate promising results for deep learning (DL) methods on a various medical images' tasks [5,7,8, 12,20]. In this paper, we aim to benchmark performance of the feature generation methods proposed by Yap et al. [21] and various DL architectures for the task of distinguishing between nuclei with and without prominent nucleolus (nuclei classification).

2   Related Work

Yap et al. [21] presented a prominent nucleoli detector along with experiments on prostate, breast, renal clear cell, and renal papillary cell cancer images. This paper proposed two gradient based methods, Histogram of Polar Gradients (HPG) and Enhanced Histogram of Polar Gradients (EHPG) and compared their detection performance with a base-line logistic regression (LR) method. HPG is essentially a SVM [3] classifier trained upon polar gradient based hand-crafted features. EHPG extended these hand-crafted features by random sampling in the gradient space and boosting [6].

There has been a recent surge of using DL for digital pathology tasks [7]. Convolutional Neural Networks (CNN) have been used for mitosis detection in breast cancer images and nuclei detection in Ki-67 stained Neuroendocrine tumor tissue images [2,18,19]. A spatially constrained CNN has been developed for nuclei detection and classification in colon cancer images [16]. Various deep CNN architectures pre-trained upon large natural image datasets have also been fine-tuned for computer-aided detection via transfer learning [15]. Stacked sparse auto-encoder network [17] has been used for nuclei detection in breast cancer images [20]. A convolutional auto-encoder [13] has been used for basal-cell carcinoma detection [4]. In this paper we benchmark the methods suggested by Yap *et al.* [21] along with basic CNN or auto-encoder based DL architectures for the task of nuclei classification. All these methods are trained from scratch for fair comparison. All the DL architectures were implemented using Caffe [10].

3   Method

3.1   Datasets

The four Hematoxylin and Eosin (H&E) stained image datasets namely, prostate, breast, renal clear cell, and renal papillary cell cancer datasets by Yap *et al.* [21] were used for our study. In these datasets, prominent nucleoli patterns have been manually annotated by a trained pathologist. All the nuclei patterns were detected using "Find Maxima" function in ImageJ [14,21]. All the nuclei without any manual annotation for prominent nucleolus pattern are defined as nuclei without any prominent nucleolus. Nuclei with and without the prominent nucleolus inside them serve as positive and negative samples respectively. The nuclei samples from all the tissue regions (cancerous and non-cancerous) are used for our study. Table 1 shows the distribution of positive and negative samples and Fig. 1 shows some example samples from each of them. Each sample is a 36×36 pixels image patch. Nuclei are stained blue in the H&E images. Nuclei without any prominent nucleolus have a uniform blue color distribution. Nuclei with a prominent nucleolus, generally have non-uniform blue color distribution with the nucleolus being stained in a darker shade of blue compared to the surrounding nucleus region stained in blue. Nuclei with prominent nucleolus are also generally larger than nuclei without them at the same magnification level.

Table 1. Positive and negative sample distribution in the four datasets.

Dataset	Number of images	Nuclei with prominent nucleolus (+ve)	Nuclei without prominent nucleolus (−ve)
Prostate cancer	5	778	922
Breast cancer	12	3753	3955
Renal clear cell cancer	9	2072	2255
Renal papillary cell cancer	9	2919	3049

3.2   Data Extraction and Augmentation

We augmented our data samples by rotation and flip operations. We rotated the original images by 90°, 180°, and 270° to generate three more images. We then flipped the original image horizontally for the fourth new image. The flipped image was also rotated 90°, 180°, and 270° to generate three more images. Hence we generated 8 times more samples when compared to the original datasets. We did our benchmark study using these augmented datasets.

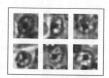

(a) **Positive samples**: Nuclei with prominent nucleolus

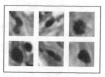

(b) **Negative samples**: Nuclei without prominent nucleolus

Fig. 1. Example positive and negative samples (36 × 36 pixels image patch) from the **prostate cancer** dataset. Nuclei with a prominent nucleolus have non-uniform blue color distribution with the nucleolus stained in a darker shade of blue compared to the surrounding nucleus stained in blue. Nuclei without any prominent nucleolus have a uniform blue color distribution.

3.3 HPG, EHPG and LR

We used the HPG, EHPG and LR methods as discussed in Yap *et al.* [21]. Each of these methods use an image patch of $s \times s$ pixels size as a input sample. The H&E images in RGB color space were converted to HSV, HLS, YCbCr, XYZ, CIELab, and CIELuv color spaces. Given a method, we then have the flexibility of selecting image patch size ($s \times s$) and color space of the input image. Yap *et al.* [21] suggested using RGB, HSV, HLS, YCbCr, XYZ, CIELab, and CIELuv color spaces and multiple image patch sizes of 12×12, 15×15, 18×18, 21×21, 24×24, 27×27, 30×30, 33×33, and 36×36. Out of all possible 63 combinations (= 7 color spaces × 9 patch sizes) we were able to use image patch sizes of 12×12, 15×15, and 36×36 along with HSV, XYZ, and CIELuv color spaces within reasonable computing time. These combinations were chosen because they performed better than other combinations in our initial experiments with LR. We generated 9 (= 3 color spaces × 3 patch sizes) weak classifiers and combined them by AdaBoost [6] for each method. All the weak classifier and boosted combination training was done using the consistent data samples across the three methods. The boosted combinations of HPG, EHPG, and LR were tested on a new set of images for the final results.

3.4 LeNet and EncoderNet

We used the CNN architecture LeNet as suggested by LeCun *et al.* [11] and implemented in Caffe [10]. It contains two successive pairs of convolutional and max-pooling layers, followed by two fully connected layers. Caffe [10]'s LeNet implementation uses Rectified Linear Units for activation instead of using Sigmoid activation as suggested in the LeCun *et al.* [11]. We also defined a stacked auto-encoder architecture, EncoderNet. It contains three fully connected hidden layers with sigmoid activation functions following the auto-encoder example implementation by Caffe [10]. Figure 2 illustrates the LeNet and EncoderNet network architectures in detail. We converted our 36×36 pixels sized three channel (RGB) image patches to grayscale and then resized them to 28×28 pixels size for use with LeNet and EncoderNet.

We modified the LeNet and EncoderNet architectures for three channel images as Color-LeNet and Color-EncoderNet (Fig. 2). In both of these networks we have the flexibility of using images from different color spaces and patch size. We generated variants of these networks by using an image patch size of 36×36 and RGB color space. We denote these variants as Color-LeNet-36 (RGB) and Color-EncoderNet-36 (RGB). We also generated boosted variants of Color-LeNet and Color-EncoderNet by using the same 9 combinations of color space and image patch size used for HPG, LR, and EHPG. These 9 combinations were then similarly boosted for final results. We also boosted all 45 weak classifiers defined for HPG, LR, EHPG, Color-Net, and EncoderNet. The neural networks were trained from scratch (random initialization). The Caffe [10] training hyper parameters were 'Base learning rate': 10^{-9} (all EncoderNet network variants), 10^{-10} (all LeNet network variants); 'Gamma': 0.0001; 'Power': 0.75; 'Momentum': 0.9 and 'Decay rate': 5×10^{-4} after every 2500 iterations. We compare ten methods for the nuclei classification task as discussed in the following section.

4   Experiments and Results

4.1   Design of Experiments

In each of the datasets we have H&E images annotated for nuclei with and without prominent nucleolus. For each annotated pixel, the square image patch around it defines the input image patch which can be categorized according to its source images. We did a three-fold cross validated study in all the datasets. The training and testing image patch sets were defined using disjoint source image sets to ensure testing results on unseen data. The training and testing data samples were same across all the ten methods for a given fold. These ten methods are LR (Boosted), HPG (Boosted), EHPG (Boosted), Color-LeNet (Boosted), Color-EncoderNet (Boosted), LeNet (Grayscale), EncoderNet (Grayscale), Color-LeNet-36 (RGB), Color-EncoderNet-36 (RGB), and 'LR, HPG, EHPG, Color-LeNet, Color-EncoderNet' (Boosted).

4.2   Results and Discussion

We report testing accuracy, F-score, and G-measure for each of the ten methods in all the four datasets. Intuitively, good classification performance would lead to higher values of these metrics. Tables 2, 3, 4, and 5 illustrate the results for prostate cancer dataset, breast cancer dataset, renal clear cell cancer dataset, and renal papillary cell cancer dataset respectively. We observe that both LR (Boosted) and Color-EncoderNet (Boosted) achieve best performance out of all individual 9 methods. 'LR,HPG, EHPG, Color-LeNet, Color-EncoderNet' (Boosted) has the best performance as per our expectations for it combines all the individual methods.

We used the nuclei classification methods as defined/suggested in the literature. The hyper parameters for HPG ($C = 2$ and $\gamma = 4$ for RBF kernel SVM)

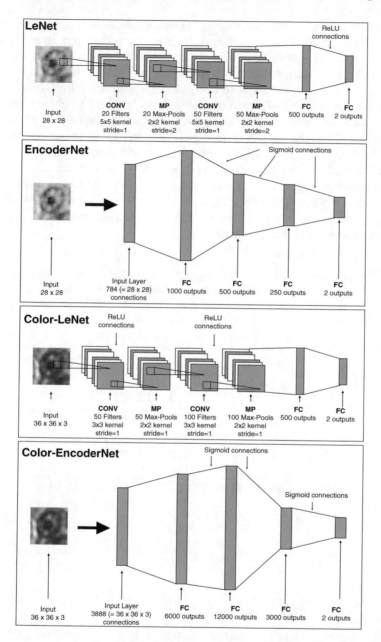

Fig. 2. Four deep learning architectures: **LeNet**, **EncoderNet**, **Color-EncoderNet**, and **Color-LeNet**. **ReLU**: Rectified Linear Unit, **CONV**: Convolution layer, **FC**: Fully connected layer, **MP**: Max-pool layer. Input for **LeNet** and **EncoderNet** is a grayscale image of size 28×28 pixels. Input for **Color-LeNet** and **Color-EncoderNet** is a three channel image of size 36×36 pixels. We derive other **Color-LeNet** and **Color-EncoderNet** variants by using an input three channel image patch of 12×12, 15×15 pixels.

and EHPG (number of randomly sampled weak classifiers is 500) were tuned in Yap *et al.* [21] for the task of distinguishing nuclei with a prominent nucleolus from stroma or cytoplasm or nuclei without prominent nucleolus. The classification task in the current study is different and harder when we are trying to distinguish between nuclei with and without prominent nucleolus. We may need further hyper parameter tuning for EHPG and HPG to obtain better performance on this harder classification task. We also observe that Color-LeNet (Boosted) and Color-EncoderNet (Boosted) perform better than their RGB variants (Color-LeNet-36 (RGB), Color-EncoderNet-36 (RGB)) and Grayscale variants (LeNet (Grayscale), EncoderNet (Grayscale)) due to their use of multiple resolution and multiple color space information. Both LeNet (Grayscale) and EncoderNet (Grayscale) use the grayscale image patches instead of three channel image patches. This would lead to some information loss and it can be one of the reasons for their low performance.

Table 2. Testing metrics for **prostate cancer** dataset. The reported values are average ± standard error across three folds. Boosted combination used image patch sizes of 12×12, 15×15, and 36×36 along with HSV, XYZ, and CIELuv color spaces. The best individual method's results are indicated in bold.

Method	Accuracy	F-score	G-measure
LR (Boosted)	**0.89 ± 0.02**	**0.85 ± 0.05**	**0.86 ± 0.04**
HPG (Boosted)	0.80 ± 0.01	0.77 + 0.02	0.77 ± 0.02
EHPG (Boosted)	0.59 ± 0.02	0.52 ± 0.05	0.55 ± 0.05
Color-LeNet (Boosted)	0.70 ± 0.03	0.66 ± 0.06	0.67 ± 0.05
Color-EncoderNet (Boosted)	0.86 ± 0.01	0.83 ± 0.02	0.84 ± 0.02
'LR, HPG, EHPG, Color-LeNet, Color-EncoderNet' (Boosted)	$0.91 \pm 4.74 \times 10^{-3}$	$0.89 \pm 5.17 \times 10^{-3}$	$0.90 \pm 5.31 \times 10^{-3}$
LeNet (Grayscale)	0.67 ± 0.05	0.60 ± 0.07	0.60 ± 0.07
EncoderNet (Grayscale)	0.76 ± 0.03	0.73 ± 0.04	0.73 ± 0.03
Color-LeNet-36 (RGB)	0.72 ± 0.03	0.69 ± 0.03	0.70 ± 0.03
Color-EncoderNet-36 (RGB)	0.55 ± 0.02	0.50 ± 0.01	0.50 ± 0.01

Table 3. Testing metrics for **breast cancer** dataset. The reported values are average ± standard error across three folds. Boosted combination used image patch sizes of 12 × 12, 15 × 15, and 36 × 36 along with HSV, XYZ, and CIELuv color spaces. The best individual method's results are indicated in bold.

Method	Accuracy	F-score	G-measure
LR (Boosted)	0.86 ± 0.02	0.83 ± 0.05	0.83 ± 0.05
HPG (Boosted)	0.85 ± 4.59 × 10^{-3}	0.84 ± 0.03	0.84 ± 0.03
EHPG (Boosted)	0.71 ± 0.03	0.72 ± 0.06	0.73 ± 0.05
Color-LeNet (Boosted)	0.78 ± 0.08	0.75 ± 0.11	0.75 ± 0.11
Color-EncoderNet (Boosted)	**0.89 ± 0.03**	**0.88 ± 0.04**	**0.88 ± 0.04**
'LR, HPG, EHPG, Color-LeNet, Color-EncoderNet' (Boosted)	0.92 ± 0.01	0.91 ± 0.03	0.91 ± 0.03
LeNet (Grayscale)	0.67 ± 0.01	0.65 ± 0.03	0.65 ± 0.04
EncoderNet (Grayscale)	0.61 ± 0.02	0.59 ± 0.03	0.59 ± 0.03
Color-LeNet-36 (RGB)	0.73 ± 0.03	0.70 ± 0.05	0.70 ± 0.05
Color-EncoderNet-36 (RGB)	0.52 ± 0.01	0.51 ± 0.01	0.52 ± 0.02

Table 4. Testing metrics for **renal clear cell cancer** dataset. The reported values are average ± standard error across three folds. Boosted combination used image patch sizes of 12 × 12, 15 × 15, and 36 × 36 along with HSV, XYZ, and CIELuv color spaces. The best individual method's results are indicated in bold.

Method	Accuracy	F-score	G-measure
LR (Boosted)	0.84 ± 0.01	0.82 ± 0.02	0.82 ± 0.02
HPG (Boosted)	0.78 ± 0.03	0.77 ± 0.03	0.77 ± 0.03
EHPG (Boosted)	0.78 ± 0.03	0.76 ± 0.02	0.77 ± 0.03
Color-LeNet (Boosted)	0.60 ± 1.37 × 10^{-3}	0.55 ± 0.02	0.56 ± 0.02
Color-EncoderNet (Boosted)	**0.88 ± 0.01**	**0.87 ± 0.01**	**0.87 ± 0.01**
'LR, HPG, EHPG, Color-LeNet, Color-EncoderNet' (Boosted)	0.89 ± 0.01	0.88 ± 0.01	0.88 ± 0.01
LeNet (Grayscale)	0.54 ± 0.02	0.49 ± 0.02	0.49 ± 0.02
EncoderNet (Grayscale)	0.65 ± 0.01	0.63 ± 0.02	0.63 ± 0.02
Color-LeNet-36 (RGB)	0.59 ± 0.03	0.57 ± 0.04	0.57 ± 0.03
Color-EncoderNet-36 (RGB)	0.53 ± 0.01	0.49 ± 0.03	0.49 ± 0.03

Table 5. Testing metrics for **renal papillary cell cancer** dataset. The reported values are average ± standard error across three folds. Boosted combination used image patch sizes of 12×12, 15×15, and 36×36 along with HSV, XYZ, and CIELuv color spaces. The best individual method's results are indicated in bold.

Method	Accuracy	F-score	G-measure
LR (Boosted)	0.89 ± 0.01	0.89 ± 0.03	0.89 ± 0.03
HPG (Boosted)	0.84 ± 0.29 × 10^{-3}	0.84 ± 0.02	0.84 ± 0.02
EHPG (Boosted)	0.81 ± 0.03	0.82 ± 0.03	0.82 ± 0.03
Color-LeNet (Boosted)	0.69 ± 0.03	0.68 ± 0.07	0.68 ± 0.07
Color-EncoderNet (Boosted)	**0.91 ± 0.01**	**0.91 ± 0.02**	**0.91 ± 0.02**
'LR, HPG, EHPG, Color-LeNet, Color-EncoderNet' (Boosted)	0.93 ± 0.01	0.93 ± 0.01	0.93 ± 0.01
LeNet(Grayscale)	0.59 ± 0.01	0.57 ± 0.04	0.58 ± 0.04
EncoderNet (Grayscale)	0.69 ± 0.02	0.69 ± 0.03	0.69 ± 0.02
Color-LeNet-36(RGB)	0.68 ± 0.02	0.66 ± 0.04	0.66 ± 0.05
Color-EncoderNet-36 (RGB)	0.55 ± 0.03	0.54 ± 2.61 × 10^{-3}	0.54 ± 4.59 × 10^{-3}

5 Conclusion

We compared nuclei classification methods trained from scratch using accuracy, F-score, and G-measure metrics. Both LR and Color-EncoderNet achieve best performance out of all nine individual methods. Future study may include pre-trained CNN and architectures with fine tuning on the histopathological datasets as suggested by Shin *et al.* [15] along with further data augmentation.

Acknowledgments. This work was supported in part by the Biomedical Research Council of A*STAR (Agency for Science, Technology and Research), Singapore; the National University of Singapore, Singapore, the Departments of Urology and Pathology at Tan Tock Seng Hospital, Singapore and Singapore-China NRF-NSFC Grant (No. NRF2016NRF-NSFC001-111). Part of the computational work for this article was done on resources of the National Supercomputing Computer Singapore (https://www.nscc.sg).

References

1. American Cancer Society: Cancer Facts & Figures 2017. American Cancer Society, Atlanta (2017)
2. Cireşan, D.C., Giusti, A., Gambardella, L.M., Schmidhuber, J.: Mitosis detection in breast cancer histology images with deep neural networks. In: Medical Image Computing and Computer-Assisted Intervention–MICCAI 2013, pp. 411–418. Springer (2013)

3. Cortes, C., Vapnik, V.: Support-vector networks. Mach. Learn. **20**(3), 273–297 (1995)
4. Cruz-Roa, A.A., Ovalle, J.E.A., Madabhushi, A., Osorio, F.A.G.: A deep learning architecture for image representation, visual interpretability and automated basal-cell carcinoma cancer detection. In: International Conference on Medical Image Computing and Computer-Assisted Intervention (MICCAI), pp. 403–410. Springer (2013)
5. Esteva, A., Kuprel, B., Novoa, R.A., Ko, J., Swetter, S.M., Blau, H.M., Thrun, S.: Dermatologist-level classification of skin cancer with deep neural networks. Nature **542**(7639), 115–118 (2017)
6. Freund, Y., Schapire, R.E.: A decision-theoretic generalization of on-line learning and an application to boosting. J. Comput. Syst. Sci. **55**(1), 119–139 (1997)
7. Greenspan, H., van Ginneken, B., Summers, R.M.: Guest editorial deep learning in medical imaging: overview and future promise of an exciting new technique. IEEE Trans. Med. Imaging **35**(5), 1153–1159 (2016)
8. Gulshan, V., Peng, L., Coram, M., Stumpe, M.C., Wu, D., Narayanaswamy, A., Venugopalan, S., Widner, K., Madams, T., Cuadros, J., et al.: Development and validation of a deep learning algorithm for detection of diabetic retinopathy in retinal fundus photographs. JAMA **316**(22), 2402–2410 (2016)
9. Irshad, H., Veillard, A., Roux, L., Racoceanu, D.: Methods for nuclei detection, segmentation, and classification in digital histopathology: a review-current status and future potential. IEEE Rev. Biomed. Eng. **7**, 97 (2014)
10. Jia, Y., Shelhamer, E., Donahue, J., Karayev, S., Long, J., Girshick, R., Guadar-rama, S., Darrell, T.: Caffe: Convolutional architecture for fast feature embedding. In: Proceedings of the 22nd ACM International Conference on Multimedia, pp. 675–678. ACM (2014)
11. LeCun, Y., Bottou, L., Bengio, Y., Haffner, P.: Gradient-based learning applied to document recognition. Proc. IEEE **86**(11), 2278–2324 (1998)
12. Madabhushi, A., Lee, G.: Image analysis and machine learning in digital pathology: challenges and opportunities. Med. Image Anal. **33**(6), 170–175 (2016)
13. Masci, J., Meier, U., Cireşan, D., Schmidhuber, J.: Stacked convolutional auto-encoders for hierarchical feature extraction. In: International Conference on Artificial Neural Networks, pp. 52–59. Springer (2011)
14. Schneider, C.A., Rasband, W.S., Eliceiri, K.W.: NIH Image to ImageJ: 25 years of image analysis. Nat. Methods **9**(7), 671 (2012)
15. Shin, H.C., Roth, H.R., Gao, M., Lu, L., Xu, Z., Nogues, I., Yao, J., Mollura, D., Summers, R.M.: Deep convolutional neural networks for computer-aided detection: CNN architectures, dataset characteristics and transfer learning. IEEE Trans. Med. Imaging **35**(5), 1285–1298 (2016)
16. Sirinukunwattana, K., Raza, S.E.A., Tsang, Y.W., Snead, D.R., Cree, I.A., Rajpoot, N.M.: Locality sensitive deep learning for detection and classification of nuclei in routine colon cancer histology images. IEEE Trans. Med. Imaging **35**(5), 1196–1206 (2016)
17. Vincent, P., Larochelle, H., Lajoie, I., Bengio, Y., Manzagol, P.A.: Stacked denoising autoencoders: learning useful representations in a deep network with a local denoising criterion. J. Mach. Learn. Res. **11**, 3371–3408 (2010)
18. Wang, H., Cruz-Roa, A., Basavanhally, A., Gilmore, H., Shih, N., Feldman, M., Tomaszewski, J., Gonzalez, F., Madabhushi, A.: Cascaded ensemble of convolutional neural networks and handcrafted features for mitosis detection. In: SPIE Medical Imaging, p. 90410B. International Society for Optics and Photonics (2014)

19. Xie, Y., Kong, X., Xing, F., Liu, F., Su, H., Yang, L.: Deep voting: a robust approach toward nucleus localization in microscopy images. In: International Conference on Medical Image Computing and Computer-Assisted Intervention (MICCAI), pp. 374–382. Springer (2015)
20. Xu, J., Xiang, L., Liu, Q., Gilmore, H., Wu, J., Tang, J., Madabhushi, A.: Stacked sparse autoencoder (SSAE) for nuclei detection on breast cancer histopathology images. IEEE Trans. Med. Imaging **35**(1), 119–130 (2016)
21. Yap, C.K., Kalaw, E.M., Singh, M., Chong, K.T., Giron, D.M., Huang, C.H., Cheng, L., Law, Y.N., Lee, H.K.: Automated image based prominent nucleoli detection. J. Pathol. Inform. **6**, 39 (2015)

Semi-automatic Segmentation of Paranasal Sinuses from CT Images Using Active Contour with Group Similarity Constraints

Zhuofu Deng[1,2(✉)], Takahiko Kitamura[3], Naoki Matsushiro[4],
Hiroshi Nishimura[3], Zhiliang Zhu[2], Min Xu[1,2], Kun Xiong[1],
and Yen-Wei Chen[1]

[1] College of Information Science and Engineering,
Ritsumeikan University, Kyoto, Japan
dengzf@swc.neu.edu.cn
[2] College of Software Engineering, Northeastern University, Shenyang, China
[3] Department of Otorhinolaryngology-Head and Neck Surgery,
National Hospital Organization Osaka National Hospital, Osaka, Japan
[4] Department of Otolaryngology, Osaka Police Hospital, Osaka, Japan

Abstract. Computerized tomographic (CT) scanning has dramatically improved the imaging of paranasal sinus anatomy as compared to sinus radiographs. Increasingly, subtle bony anatomic variations and mucosal abnormalities of this region are being detected. The morphological knowledge of nasal cavity and paranasal sinuses has an important clinical value. It is used for the detection of sinus pathologies, for determination of therapy, planning of endoscopic surgeries and for surgical simulations. Current research and industry assisting systems need a workspace definition of the paranasal sinuses, which is realized by segmentation. This paper presents a semi-automatic segmentation method for the paranasal sinuses which allows us to locate structures. In general, the traditional active contour methods like Snake, Levelset can resolve the CT images of paranasal sinuses normal without any anatomic variations caused by sinusitis. However, in the clinical practice, the diseased radiological image has more significances so that these classical methods can not work satisfied very well as the boundaries of sinuses has been covered by impurity inflammation produced. At this point, we proposed a novel method group similarity based on Low Rank to repair the lost part of the boundary. The experiment results proved that our proposed method outperformed conventional algorithms especially in abnormal images.

1 Introduction

There are many variations in nasal and paranasal anatomy. CT scans are useful tools for understanding of complicated structure of this anatomy. We can get the 3-D dimensional image from coronal, axial and sagittal slice CT images. Coronal plane computerized tomographic is a promising modality for capturing images of the human upper airway, where More and more attentions have been payed.

© Springer International Publishing AG 2018
Y.-W. Chen et al. (eds.), *Innovation in Medicine and Healthcare 2017*, Smart Innovation,
Systems and Technologies 71, DOI 10.1007/978-3-319-59397-5_10

The complicated structure of paranasal sinuses in images have been used to visualize and analyze treatment efficiency in subjects with chronic sinusitis and other sinus diseases [1,2]. Providing accurate modalities for morphofunctional analysis are essential for improving diagnosis, treatment planning, and assessing treatment outcomes [3]. Therefore, segmentation this medical image processing technique should be indispensable, and improved data acquisition techniques with higher resolution deliver new possibilities for medical image segmentation [26]. Recently, three-dimensional model data of nasal cavity and paranasal sinuses is useful in many clinical applications.

Surgical plan: endoscopic sinus surgery of the chronic rhinosinusitis needs an accurate evaluation of diseases and paranasal anatomic variations [11]. Many anatomic variations of the structures in the middle meatus can narrow the stenotic clefts even more, and thus predispose to more or less intense contact of apposing mucosal surfaces. This may impede or block ventilation and drainage of the ethmoid and surrounding larger sinuses and thus affect them as well. After identification of these variations, functional endoscopic sinus surgery with usually minimal invasive operations often can provide dramatic relief of chronic sinusitis symptoms, particularly headache that may have been present for months or even years [21]. In other words, it is essential that the surgeon has exact knowledge of the anatomy and anatomic variations of nasal cavity and paranasal sinuses. Anatomical variations of paranasal sinuses can be distinct for different patients, as it can be seen by the number of ethmoidal cells, ranging from 3 to 18 per side [14,17].

Volume quantification: volume quantification has been most important in the evaluation of the paranasal sinus. Comparing with other body size indices, the increment of the size was quite apparent and the lower prevalence of sinusitis was assumed to play a role in this increased volume of paranasal sinuses in the modern population [13]. Although there are some uncertain relations in this idea, the volume measurement has been another approach for sinusitis diagnosis.

2   Related Work

The segmentation ways of paranasal sinuses are divided into two categories: manual and automatic segmentation. Automatic segmentation of medical images is important of several reasons. Manual segmentation is tedious and not scalable to large datasets which will bring the large of cost patients and radiologists. Automatic segmentation allows us to extract shapes from a large number of large populations. These shapes or structures play an important role in registration [20]. Some algorithms have been proposed in recent years for this problem. Zein Salah et al. [19] introduced an algorithm based on Region Growing techniques and its merits include computing fast. Nevertheless, the complexity in intensity of paranasal sinuses especially in boundary impact the results of this sort of algorithms depending on gray scale of images severely. In some other semi-automatic algorithms like [7,15], although they adopt different thoughts, they are all related to intensity of pixels and certainly they can not not resolve the vague boundary problem sometimes which can not be observed by eyes, for example as depicted Fig. 1 with red circles. This problem is due to the patients'

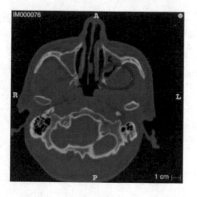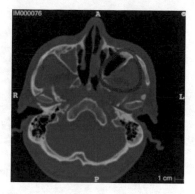

Fig. 1. Vague boundaries in paranasal sinuses

nasal diseases. Accordingly, the algorithms can not find the entire boundary and will give the unacceptable results.

In this paper, we propose a novel method combining traditional segmentation algorithm active contour based on Snake and group similarity. The active contour method is responsible for generating initial contour and the group similarity is for the constraints of evolving in curve. Group similarity came from Low Rank idea and it can keep boundaries shapes uniform. Inspired by this point, we can make use of this constraint to limit slices each other in order that the slices of lost boundaries can fix them correctly and automatically. The experiment result showed that our proposed method can repair the abnormal contours and could be accepted by the radiologists.

3   Method

3.1   Overview of the Proposed Method

The proposed method consists of two parts that include active contour based on snake [6] and group similarity [29]. Figure 2 presents an overview of the proposed algorithm. In this work, at the beginning, initial segmentation contains some tasks about image preprocessing for example Gauss smoothing. After that with the initial contour by users themselves, the system will create an active contour mathematical model for iterations. When the users set the appropriate step length, the algorithm will run until convergence. When it attain some times later, the algorithm will stop for checking whether the some slices have the similar contours. If not the system will correct the abnormal ones and continue the active contour iterations. We will repeat this process before the model's energy has no change.

3.2   Active Contour

Active contour methods offer several advantages. First, an active contour is modeled directly as a curve and is maintained as a curve throughout the iterative

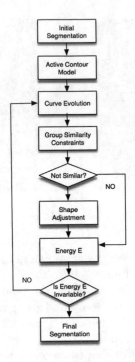

Fig. 2. The flowchart of the proposed segmentation method

process that deforms it toward the final solution. Thus, characteristics of the desired curve such as its length, curvature, and conformation to the data can be evaluated or imposed as an explicit part of the algorithm. Second, an optimality criterion involving both the data is specified, and an optimal solution is sought. Third, an explicit map between the curve and the unit interval is generated automatically. Such a map can be used to determine properties of the cortex such as lengths, tangents, normals, and curvature using the theory of differential geometry. A final benefit of this approach is that it can be readily extended to 3-D by defining deformable surfaces instead of curves. Some work along these lines has already been reported in [8,9,23,24].

3.3 Group Similarity Constraints

In real applications, the performance of the active contour model is prone to be corrupted by missing and misleading features especially in the paranasal sinuses CT images with the disturbing of opacitas. For example, segmentation of the pairs of maxillary sinuses is still an unresolved problem due to the characteristic artefacts as the patients with sever sinusitis or other nasal disease. The normal people's sinuses are filled with pure air so that you can see nothing but black in the images. However, if the patients have nasal inflammation the space of sinuses are presenting some low gray scale color and because the sinuses boundary consist of bones which shows white color, the contours will be disturbed. To improve the robustness of active contours, the shape prior is often used. The prior knowledge of the shape to be segmented is modeled based on a set of manually-annotated shapes to guide the segmentation. Previous deformable template models [10,12,16,28] can be regarded as the early efforts towards knowledge-based segmentation. Recently, most of papers about active contour focus on constructing an energy equation for shape constraints, and in fact they always use distinctions between two areas in the contour and out of the contour [22,25,27]. Due to the complexity of gray scale in the paranasal sinuses images these algorithm all fails to have a perfect segmentation result and always you will find a leakage of sinuses boundary.

In this paper we propose a novel algorithm which is referred to [29] and improved for fitting maxillary sinuses segmentation. At first, given a sequence of images $I_1, ..., I_n$, we try to find a set of contours $I_1, ..., I_n$ to segment the object in these images. To keep the contours similar to each other, we propose to segment the images by

$$\min_X \sum_{i=1}^{n} f_i(C_i), \quad \text{subject to } \text{rank}(X) \leq K \qquad (1)$$

where $X = [C_1, ..., C_n]$ and K is a predefined constant. $f_i(C_i)$ is the energy of an active contour model to evolve the contour in each frame such as snake, geodesic active contour and region-based models. Since rank is a discrete operator which is both difficult to optimize and too rigid as a regularization method, we propose to use the following relaxed form as the objective function:

$$\min_X \sum_{i=1}^{n} f_i(C_i) + \lambda \|X\|, \quad \text{subject to } \text{rank}(X) \leq K \qquad (2)$$

Here, $rank(X)$ is replace by the nuclear norm $\|X\|_*$, i.e. the sum of singular values of X. Recently, the nuclear norm minimization has been widely used in low-rank modeling such as matrix completion and robust principal component analysis. As a tight convex surrogate to the rank operator, the nuclear norm has several good properties: Firstly, the convexity of the nuclear norm makes it possible to develop fast and convergent algorithms in optimization. Secondly, the nuclear norm is a continuous function, which is important for a good regularize in many applications. For instance, in our problem the small perturbation in the shapes may result in a large increase of $rank(X)$, while $\|X\|_*$ may rarely change.

According to [29], we use to Proximal Gradient (PG) method [4,18] this problem where $rank(X)$ is non-convex function so we change the minimization of nuclear in the contours group position matrix. Distinctly, in our algorithm we change the contour similarity design. Because in our work, we intend to process images in 2D condition and repair the lost edges. At first, in the slices we will find three images in good quality in the left, middle and right positions of image series where boundaries of sinuses are clear. In addition, we will mark the entire contours in these three slices as the key. So, after we build the similarity matrix we select them as the fixed vectors, we minimize the objective energy functions with PG method in iterations for getting acceptable results. If the abnormal anatomical or something happened, this algorithm will fix the problems. The intuition of our algorithm is that, at each iteration, we first evolve the active contours according to the image based forces and then impose the group similarity regularization via singular value thresholding.

4   Experimental Results

4.1   Data

In this work, 30 paranasal sinuses volumes of CT images from JCHO Osaka hospital were used in the experiments. 40 percents of subjects were scanned with SIEMENS SOMATAM Definition AS+, and the remaining is from GE MEDICAL SYSTEMS Discovery CT750 HD. The images' exposure time from GE is surrounding 1460 ms and SIEMENS is about 1000 ms. The data pixel spacing and slice thickness in SIEMES are 0.390625 mm and 0.6 mm, whereas the other one are 0.351562 mm and 0.625 mm. These data set allow us to test the robustness of the algorithm. At the beginning of the experiment, an experienced radiologists will check whether the data loss important information about the bone boundary around maxillary sinuses which can be not segmented by automatic or semi-automatic software he has used. To avoid the situation that radiologist remembered previous manual segmentation, any two manual segmentations were performed at least one week apart. The segmentation result from our algorithm was compared with other software's and manual's respectively. Two experienced experts will give the conclusion whether our proposed segmentation can get a accepted effect.

4.2   Implementation Details

The main algorithm was implemented in MATLAB codes. The algorithm ran on Mac Sierra OS 10.12.2 with an 2.5 GHz Intel Core i7 and 16 GB 1600 MHz DDR3. Our code is not optimized and does not use multi-thread, GPU acceleration or parallel programming. In order to improve our algorithm speed, we optimize a little part of the method where if we set the active contour step 1, every iteration the contour similar will not change too much, so it will cost lots of time in computing. Therefore, we reset our active contour iterations that after 20 steps of it we let the algorithm check the similarity of the slices which will save more time for significant computing.

4.3   Qualitative Results

Among the testing data, the expert has chosen the complicated images for segmentation. For example Fig. 1, in the red circle the bone loss essential informations so common semi-automatic can not get good result in Fig. 3(b). Figure 3(a)

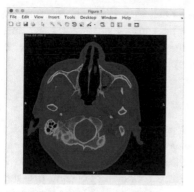

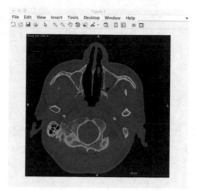

(a) Initial contour by users in slice   (b) Only active contour fails in slice

Fig. 3. Test with common software algorithm

is the initial contour user drew at will but the circle must be ensured within the maxillary sinus. This initialization requirement is the same in our proposed algorithm.

For easy computing and observations, after we select three key slices, we extracted the bones in the CT images and threshold the images into 0 and 1. So, the supposed algorithm will process the successive images according to key slices. Figure 4 shows the experiment's result. As observed, the boundaries which loss the important information has been filled up have been accept by

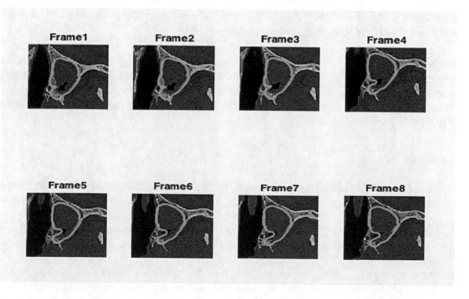

Fig. 4. The flowchart of the proposed segmentation method

the radiologist. Due to the insufficient groups of datasets, we can not gather the statistics to have a conclusion. Though six group abnormal images we find our proposed algorithm accounts for more 20 percents of time that the one with not group similarity constraints, so it is well acceptable.

5   Conclusion

Our method combined active contour and group similarity for paranasal sinuses segmentation has resolve a big problem that some boundaries are not complete in the CT images some nasal diseases result in. At some points it can bring with new possibilities for patients' data volume rendering so that pre-surgical plan and sinuses volume measurement will go on smoothly. There are also some limitations of this algorithm:

1. The key slices: How to choose is a hinge. If choose the unacceptable boundary will be dangerous in the segmentation which needs more experience.
2. What it abnormal: In some patients, the nasal bone sometimes and somewhere have lots of variations [5]. So it is difficult to judge if the shape translation is born or lost informations lead to.
3. λ choice: In the group similarity energy equation there is an experienced parameter controlling the weight of similarity in the curve evolution. If it is too large, the evolution will stop and can not attain the actual boundary. And if it is too small, the group similarity will have not impacts on the segmentation. So what is proper will be next job in the future.

In the future, we will collect large scale of the data for testing in the quality and quantity of time. Improving our algorithm for more complicated environment is necessary. At the same time, more accurate 3D segmentation algorithm with group similarity is also certainly our future work.

Acknowledgments. This research was supported in part by the Grant-in Aid for Scientific Research from the Japanese Ministry for Education, Science, Culture and Sports (MEXT) under the Grant No. 15H01130 and No. 16H01436, in part by the MEXT Support Program for the Strategic Research Foundation at Private Universities (2013–2017).

References

1. Alsufyani, N., Flores-Mir, C., Major, P.: Three-dimensional segmentation of the upper airway using cone beam CT: A systematic review. Dentomaxillofacial Radiology (2014)
2. Alsufyani, N.A., Al-Saleh, M.A., Major, P.W.: CBCT assessment of upper airway changes and treatment outcomes of obstructive sleep apnoea: A systematic review. Sleep Breathing **17**(3), 911–923 (2013)
3. Alsufyani, N.A., Hess, A., Noga, M., Ray, N., Al-Saleh, M.A., Lagravère, M.O., Major, P.W.: New algorithm for semiautomatic segmentation of nasal cavity and pharyngeal airway in comparison with manual segmentation using cone-beam computed tomography. Am. J. Orthod. Dentofac. Orthop. **150**(4), 703–712 (2016)

4. Beck, A., Teboulle, M.: A fast iterative shrinkage-thresholding algorithm for linear inverse problems. SIAM J. Imaging Sci. **2**(1), 183–202 (2009)
5. Bolger, W.E., Parsons, D.S., Butzin, C.A.: Paranasal sinus bony anatomic variations and mucosal abnormalities: CT analysis for endoscopic sinus surgery. Laryngoscope **101**(1), 56–64 (1991)
6. Bresson, X., Esedoglu, S., Vandergheynst, P., Thiran, J.P., Osher, S.: Fast global minimization of the active contour/snake model. J. Math. Imaging Vis. **28**(2), 151–167 (2007)
7. Bui, N.L., Ong, S.H., Foong, K.W.C.: Automatic segmentation of the nasal cavity and paranasal sinuses from cone-beam CT images. Int. J. Comput. Assist. Radiol. Surg. **10**(8), 1269–1277 (2015)
8. Cohen, I., Cohen, L.D., Ayache, N.: Introducing new deformable surfaces to segment 3D images. In: 1991 Proceedings of the IEEE Computer Society Conference on Computer Vision and Pattern Recognition, CVPR 1991, pp. 738–739. IEEE (1991)
9. Cohen, L.D., Cohen, I.: A finite element method applied to new active contour models and 3D reconstruction from cross sections. In: 1990 Proceedings of the Third International Conference on Computer Vision, pp. 587–591. IEEE (1990)
10. Deformable, P.: Boundary finding with parametrically deformable models. IEEE Trans. Pattern Anal. Mach. Intell. **14**(11), 1061 (1992)
11. Ferrie, J., Azais, O., Vandermarcq, P., Klossek, J., Drouineau, J., Gasquet, C.: X-ray computed tomographic study of the ethmoid and middle meatus. II. Radioanatomy (axial incidence) and morphological variations. J. de radiologie **72**(10), 477–487 (1991)
12. Jain, A.K., Zhong, Y., Lakshmanan, S.: Object matching using deformable templates. IEEE Trans. Pattern Anal. Mach. Intell. **18**(3), 267–278 (1996)
13. Kawarai, Y., Fukushima, K., Ogawa, T., Nishizaki, K., Gunduz, M., Fujimoto, M., Masuda, Y.: Volume quantification of healthy paranasal cavity by three-dimensional CT imaging. Acta Otolaryngol. **119**(540), 45–49 (1999)
14. Lang, J.: Clinical Anatomy of the Nose, Nasal Cavity and Paranasal Sinuses. Thieme, New York (1989)
15. Li, S., Fevens, T., Krzyżak, A.: A SVM-based framework for autonomous volumetric medical image segmentation using hierarchical and coupled level sets. Int. Congr. Ser. **1268**, 207–212 (2004). Elsevier
16. Metaxas, D.N.: Physics-Based Deformable Models: Applications to Computer Vision, Graphics and Medical Imaging, vol. 389. Springer, Heidelberg (2012)
17. Navarro, J.A.C., de Lima Navarro, J., de Lima Navarro, P.: The Nasal Cavity and Paranasal Sinuses Surgical Anatomy. Springer, Heidelberg (2001)
18. Nesterov, Y., et al.: Gradient methods for minimizing composite objective function (2007)
19. Salah, Z., Bartz, D., Dammann, F., Schwaderer, E., Maassen, M.M., Straßer, W.: A fast and accurate approach for the segmentation of the paranasal sinus. In: Meinzer, H.P., Handels, H., Horsch, A., Tolxdorff, T. (eds.) Bildverarbeitung für die Medizin 2005, pp. 93–97. Springer, Heidelberg (2005)
20. Sinha, A., Leonard, S., Reiter, A., Ishii, M., Taylor, R.H., Hager, G.D.: Automatic segmentation and statistical shape modeling of the paranasal sinuses to estimate natural variations. In: SPIE Medical Imaging, p. 97840D. International Society for Optics and Photonics (2016)
21. Stammberger, H., Wolf, G.: Headaches and sinus disease: The endoscopic approach. Ann. Otol. Rhinol. Laryngol. **97**(5), 3–23 (1988)

22. Tang, Y., Li, X., von Freyberg, A., Goch, G.: Automatic segmentation of the papilla in a fundus image based on the CV model and a shape restraint. In: 2006 18th International Conference on Pattern Recognition, ICPR 2006, vol. 1, pp. 183–186. IEEE (2006)

23. Terzopoulos, D.: The computation of visible-surface representations. IEEE Trans. Pattern Anal. Mach. Intell. **10**(4), 417–438 (1988)

24. Terzopoulos, D., Waters, K.: Physically-based facial modelling, analysis, and animation. J. Visual. Comput. Animation **1**(2), 73–80 (1990)

25. Tian, Z., Liu, L., Zhang, Z., Fei, B.: Superpixel-based segmentation for 3D prostate MR images. IEEE Trans. Med. Imaging **35**(3), 791–801 (2016)

26. Tingelhoff, K., Eichhorn, K.W., Wagner, I., Kunkel, M.E., Moral, A.I., Rilk, M.E., Wahl, F.M., Bootz, F.: Analysis of manual segmentation in paranasal CT images. Eur. Arch. Otorhinolaryngol. **265**(9), 1061–1070 (2008)

27. Wang, L., Li, C., Sun, Q., Xia, D., Kao, C.Y.: Active contours driven by local and global intensity fitting energy with application to brain MR image segmentation. Comput. Med. Imaging Graph. **33**(7), 520–531 (2009)

28. Yuille, A.L., Hallinan, P.W., Cohen, D.S.: Feature extraction from faces using deformable templates. Int. J. Comput. Vis. **8**(2), 99–111 (1992)

29. Zhou, X., Huang, X., Duncan, J.S., Yu, W.: Active contours with group similarity. In: Proceedings of the IEEE Conference on Computer Vision and Pattern Recognition, pp. 2969–2976 (2013)

Advanced ICT for Medicine and Healthcare

The Possibility of Hemorheological Parameters as Precursors of Recurrent Strokes

Margarita V. Kruchinina[1](✉), Andrey A. Gromov[1],
Vladimir M. Generalov[2], Vladimir N. Kruchinin[3],
and Gennadiy V. Shuvalov[4]

[1] Federal State Budgetary of Scientific Institution "Institution of Internal
and Preventive Medicine", Novosibirsk, Russia
kruchmargo@yandex.ru
[2] Federal Budget Institution of Science State Research Center of Virology
and Biotechnology "Vector", Novosibirsk Region, Koltsovo, Russia
[3] Rzhanov Institute of Semiconductor Physics, Siberian Branch Russian
Academy of Science, Novosibirsk, Russia
[4] Federal State Unitary Enterprise Siberian Scientific Research Institute
of Metrology, Novosibirsk, Russia

Abstract. The study aimed to evaluate the possibility of application of hemorheological parameters as precursors of recurrent strokes in cases of various pathogenetic forms of the disease.

214 patients with acute disturbances in cerebral circulation were included into the study. Their mean age was 47,7 + 5,6 years. 197 of them suffered from ischemic and 17 - from hemorrhagic stroke at its sub-acute or residual stage. 162 patients were studied within the course of administered therapy. Control comprised 35 healthy subjects of similar age.

Electric and viscoelastic behavior of erythrocytes was studied using dielectrophoresis using an electrooptical cell detection system. The content of total lipids and composition in erythrocyte membranes were investigated by two-dimensional thin layer chromatography using plates Kissellgell "Merck" F254 (Germany). We applied routine methods to study parameters of hemostasis, genetic mutations and activity rate of enzymes of erythrocytes.

The revealed various pathogenetic variants of stroke require different emphasis to be given to the course of therapy. We came up with list of hemorheological parameters (electrical, viscoelastic behavior of erythrocytes, indices of hemostasis) that serve as precursors of recurrent stroke.

Keywords: Recurrent stroke · Precursor · Hemorheological parameters · Dielectrophoresis

1 Introduction

According to the World Health Organization, 15 million people suffer stroke worldwide each year. Of these, 5 million die and another 5 million are permanently disabled. Stroke is the third leading cause of death in the United States. More than 140,000 people die each year from stroke in the United States. Each year, approximately 795,000 people

© Springer International Publishing AG 2018
Y.-W. Chen et al. (eds.), *Innovation in Medicine and Healthcare 2017*, Smart Innovation, Systems and Technologies 71, DOI 10.1007/978-3-319-59397-5_11

suffer a stroke. About 600,000 of these are first attacks, and 185,000 are recurrent attacks [1]. Despite the existing approaches to the diagnosis of stroke [2–5], the development of the new predictors of recurrent cerebrovascular events remains an actual problem.

The study aimed to evaluate the possibility of application of hemorheological parameters as precursors of recurrent strokes in cases of various pathogenetic forms of the disease.

2   Materials and Methods

214 patients with acute disturbances in cerebral circulation were included into the study. Their mean age was 47,7 $\pm$ 5,6 years. 197 of them suffered from ischemic and 17 – from hemorrhagic stroke at its sub-acute or residual stage. 162 patients were studied within the course of administered therapy. Control comprised 35 healthy subjects of similar age.

Upon the initial examination the patients were divided into the two study groups depending upon viscoelastic and electric parameters of erythrocytes. In other words, we revealed the different hemorheological variants of a stroke: 149 patients had the so-called "hard erythrocytes" at the background of metabolic syndrome (Group I); 65 patients had "fragile" cells at the background of dysplasia of connective tissue, viral infections, and disturbances in liver function without traditional risk factors (p < 0,001) (Group II).

Electric and viscoelastic behavior of erythrocytes was studied using dielectrophoresis using an electrooptical cell detection system [6].

The content of total lipids (g/l) and composition (in%) in erythrocyte membranes were investigated by two-dimensional thin layer chromatography using plates *Kissellgell "Merck" F_{254}* (Germany) [7]. We applied routine methods to study parameters of hemostasis, genetic mutations and activity rate of enzymes of erythrocytes.

The Fig. 1 shows a measurement cell used in dielectrophoresis.

The essence of the method lies in the fact that the individual interaction of a suspension cell with non-uniform alternating electric field is accompanied by polarization of its electric charges [8]. At standard parameters of the field the newly-formed dipole moment depends upon the parameters of the cell [9]. We observed in the electric filed the effects as follows (Fig. 2):

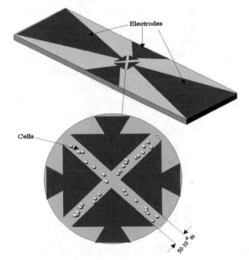

Fig. 1. Design of the measuring device for dielectrophoresis.

- directional transport and sedimentation of cells on electrodes,
- cells get accumulated along the electric field line,
- deformation of a cellular shape,
- rotation of single cells,
- destruction, etc.

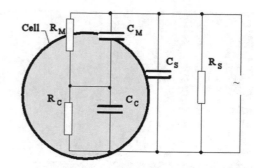

Both external and internal structures of cells can be selectively studied based on changes in generator frequency. At low frequencies we could study electric parameters of the cellular membrane and the surface of cells while at higher frequencies we could evaluate electric parameters of its cytoplasm [10, 11].

Fig. 2. The electrical equivalent schema of the cells in an alternating electric field: R_M, C_M - the active resistance and capacitance of the cell membrane; R_C, C_C - the active resistance and capacitance of the cell cytoplasm; R_S, C_S - the active resistance and capacitance of the cell surface.

This is the way erythrocytes look like when applying dielectrophoresis (Fig. 3).

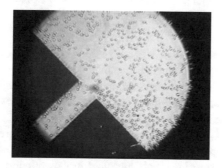

Fig. 3. Erythrocytes in non-uniform alternating electric field (NUAEF) using the method of dielectrophoresis.

3   Results

The Fig. 4 shows the samples of behavior of erythrocytes in patients of various study groups. Erythrocytes of healthy subjects tend to have high amplitude of deformation and the pace of shift towards electrodes. Plasticity of red blood cells is significantly decreased in patients of Group I (p < 0,01). In patients of Group II deformability of erythrocytes was moderately reduced. However, the cells possessed high potential to aggregate and destruct (p = 0,02) (Fig. 4 C, D).

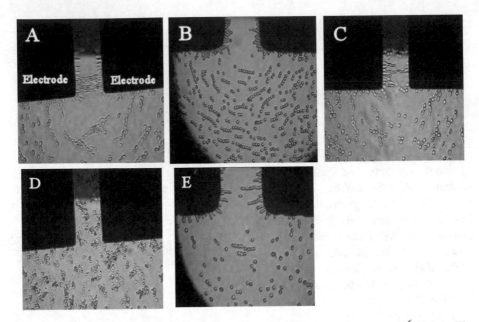

Fig. 4. Examples of behaviour of erythrocytes in NUAEF at the frequency 10^6 Hz: A. The control group; B. Group I – low amplitude of deformation of red blood cells; C. Group II - high index of erythrocyte destruction; D. Group II - high index of erythrocyte aggregation; E. Group II - moderately decreased amplitude of erythrocyte deformation.

Differences in electric and viscoelastic behavior of erythrocytes correlated with the level of lipids in cellular membranes. It means that in Group I we observed significantly higher fractions of total lipids, cholesterol and fraction of triglycerides ($p = 0,03–0,05$), while the content of total phospholipids and cholesterol ethers were lower as compared to Group II and control ($p < 0,001−0,05$) (Fig. 5).

Changes in hemostasis speak in favor of the possible presence of pathogenetic differences among the groups of patients under study. Patients of Group I demonstrated activation of cellular hemostasis and endothelium dysfunction ($p < 0,05$) (Table 1). Patients of Group II were revealed to have disturbances in cellular he-mostasis, thrombocytopathy, i.e. aggregation-related activity of platelets was decreased ($p = 0,04$); the reaction of irritation of immune system (i.e. low level of leukocyte-and-platelet aggregation, $p < 0,05$) was present. In cases of stroke con-sumption of fibrinolysis factors speaks in favor of a more pronounced hemorrhagic seepage ($p < 0,05$).

Based on the data obtained different emphasis was given to the courses of the administered therapy. In Group I patients with "rigid" erythrocytes were administered essential phospholipids, antioxidants and the preparations increasing the level of intracellular macroergic compounds in addition to the routine complex of antithrombotic and thrombolytic therapy. Positive dynamic changes in viscoelastic parameters of erythrocytes were confirmed by the significant increase of amplitude of cellular defor-mation, capacity, pace of shift of erythrocytes towards electrodes, the value of a dipole

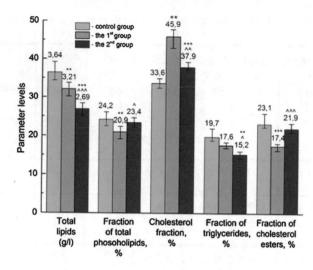

Fig. 5. The content of total lipids (g/l) and composition (in %) in erythrocyte membranes, X + m (*Note: * the significance of differences from the control group: *-p < 0,05; **-p < 0,01; ***-p < 0,001 ^ the significance of differences from Group I: ^-p < 0,05; ^^-p < 0,01; ^^^-p < 0,001.*)

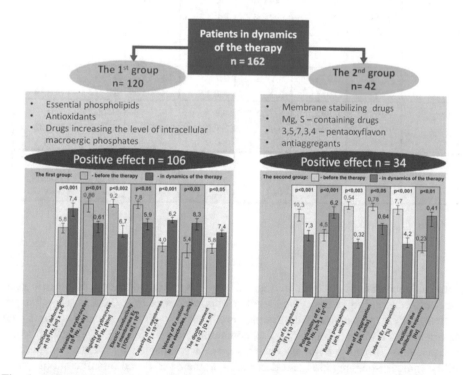

Fig. 6. Different accents in the administered therapy of patients with stroke and changes the parameters of red blood cells in the dynamics of treatment.

Table 1. Parameters of hemostasis in patients with stroke and in the control group

Groups	Degree of platelet aggregation with, %				Hemo-lysate – aggre-gation test x10⁻², s.	Soluble fibrin monomer complex, g/l	Hageman dependent fibrino-lysis, min.	Index inactivation of trombin
	Ricto-mycin 1,0	Colla-gen	ADP 5,0	ADP 0,5				
The Controls	65,4 ± 5,27	58,92 ± 8,84	62,95 ± 7,54	11,07 ± 5,32	15,17 ± 0,22	0,027 ± 0,018	10,21 ± 3,17	1,8 ± 0,17
Group I	51,48 ± 3,0 *	55,27 ± 2,88	43,13 ± 9,15 *	8,74 ± 2,31	13,6 ± 0,18 *	0,088 ± 0,002 ***	28,12 ± 3,89 ***	2,18 ± 0,03 *
Group II	40,32 ± 3,86 **^	34,64 ± 4,57 **^^	37,23 ± 7,8 **^	6,42 ± 2,43 *	14,56 ± 0,13 *^	0,062 ± 0,001 **^^	18,15 ± 2,25 **^^	2,34± 0,07 **

Groups	Von Wile-brand factor, %	Lupus anticoa-gulant, cond. U	Rate of leukcyte-platelet aggregation, s.	Fibrinogen, g/l	Anti-thrombin, %	Protein C, %	d-Dimer, ng/ml
The Controls	94,12 ± 7,73	0,92 ± 0,035	8,12 ± 0,069	2,1 ± 0,12	123,4 ± 7,2	112,4 ± 3,8	0,16 ± 0,13
Group I	140,6 ± 9,2 **	0,99 ± 0,005 *	8,97 ± 0,02 *	4,1 ± 0,07 **	84,5 ± 4,1 *	93,58 ± 2,46 *	0,94 ± 0,32 ***
Group II	114,3 ± 8,0 *^	1,15 ± 0,01 **^	11,3 ± 0,04 *^	5,8 ± 0,14 *^^	77,2 ± 3,4 **^	75,2 ± 4,3 **^	0,59 ± 0,25 **^

Note as in Fig. 5

moment, polarizability at high frequencies and significant decrease of electric conductivity, generalized viscosity and rigidity of erythrocytes (p < 0,001−0,05) (Fig. 6).

In Group II the patients with "fragile" erythrocytes were additionally administered membrane stabilizing drugs, Mg-, S-containing drugs, 3,5,7,3',4'-pentaoxiflavon, antiaggregants. Positive dynamic changes were accompanied by increase of polarizability, indices of aggregation and destruction at the background of sufficient plasticity of cells (p < 0,001−0,05).

Positive shifts in parameters of erythrocytes correlated with the "+" dynamics of MRT and CT (r = 0,88, p < 0,01), decrease of severity of neurologic symptoms (r = –0,72, p = 0,03), positive shifts in parameters of hemostasis (r = 0,69, p = 0,02).

The absence of positive dynamic changes or negative dynamics of erythrocyte parameters (low deformity rate of erythrocytes, their high potential to aggregate and hemolysis, low capacity, dipole moment, polarizability) in combination with the increased levels of hematocrit, low indices of aggregation activity of platelets, levels of protein C and high levels of D-dimer, soluble fibrin monomer complex (Fig. 7) within the course of studies of patients (14 subjects of Group I and 11 patients of Group II) correlated with the recurrent ischemic cerebral attacks (r = 0,67, p = 0,04), increase of MRT-registered areas of hypoxia (r = 0,44, p < 0,05), aggravation of neurologic symptoms (r = 0,59, p < 0,01). These parameters of erythrocytes and hemostasis were named by us the "alert markers".

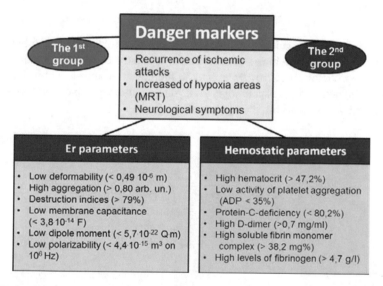

Fig. 7. The parameters of erythrocytes and hemostasis and their levels - precursors of recurrent strokes.

ROC curves were plotted for dielectrophoresis and parameters of hemostasis to demonstrate the ability to predict the recurrent strokes (Fig. 8). The area under the ROC curve (AUROC) was from 0,719 to 0.831 for the various red blood cell parameters and from 0,745 to 0,868 for the various hemostasis parameters. Values greater than 0,8 indicate excellent predictive ability [12].

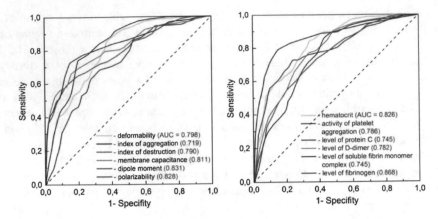

Fig. 8. ROC curve analysis of parameters of hemostasis and erythrocytes as predictors of recurrent strokes.

The majority of the patients with negative dynamics (68%) tended to have hetero- or homozygous gene mutations of the system of hemostasis, i.e. mutations in prothrombin gene, factor V Leiden as well as in the genes of anti-thrombin III, C and S proteins, tissue activator of plasminogen, enzymes of homocysteine metabolism, which is confirmed by other authors [3, 4] (Fig. 9).

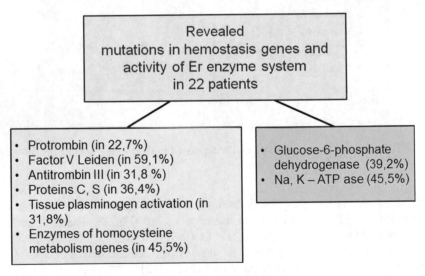

Fig. 9. Revealed mutations in the genes of hemostasis and activity of erythrocyte enzyme system.

Obviously, those patients had associated mutations of enzymatic system of ery-throcytes and carcass albumins of erythrocytic membranes [13]. This, to a certain degree, reflects changes in activity rate of glucose-6-phospahte-dehydrogenase, K+ , Na+ − Mg2+ -dependent ATPase of erythrocytes (p < 0,01–0,05).

Such defects can possibly serve as one of the causes of unsuccessful therapy within the course of study [6]. However, further studies are required and other medical preparations are to be administered.

4 Conclusions

The revealed various pathogenetic variants of stroke require different emphasis to be given to the course of therapy. We came up with list of hemorheological parameters (electrical, viscoelastic behavior of erythrocytes, indices of hemostasis) that serve as precursors of recurrent stroke.

References

1. World Health Organization: International Classification of Diseases, 10th Re-vision. apps. who.int/classifications/apps/icd/icd10online/. Accessed 21 Aug 2009
2. Cole, R., et al.: Hemostatic markers in patients at risk of cerebral ischemia. Stroke **31**(8), 1856–1862 (2000)
3. Hosomi, N., et al.: Predictors of intracerebral hemorrhage severity and its outcome in Japanese stroke patients. Cerebrovasc. Dis. **27**, 67–74 (2008)
4. Hanger, H.C., et al.: The risk of recurrent stroke after intracerebral haemorrhage. J. Neurol. Neurosurg. Psychiatry **78**, 836–840 (2007)
5. Eriksson, S.E., Olsson, J.E.: Survival and recurrent strokes in patients with different subtypes of stroke: a fourteen-year follow-up study. Cerebrovasc. Dis. **12**, 171–180 (2001)
6. Generalov, V.M., et al.: Dielectrophoresis in the Diagnosis of Infectious and Noninfectious Diseases, Novosibirsk-Tseris (2011). 172 p.
7. Kates, M.: Techniques of Lipidology, Isolation, Analysis and Identification of Lipids. Elsevier Publishing Co., Inc., New York (1972). 322 p.
8. Pohl, H.A.: Dielectrophoresis. Cambridge University Press, Cambridge (1978). 579 p.
9. Jones, T.B.: Basic theory of dielectrophoresis and electrorotation. IEEE Eng. Med. Biol. Mag. **22**(6), 33–42 (2003)
10. Archer, S., et al.: Cell reactions to dielectrophoretic manipulation. Biochem. Biophys. Res. Commun. **257**, 687–698 (1999)
11. Burt, J.P., et al.: Dielectrophoretic characterization of Friend murine erythroleukaemic cells as a measure of induced differentiation. Biochim. Biophys. Acta. **1034**, 93–101 (1990)
12. Grinhald, T.: Evidence-based medicine. GEOTARG – Media, Moscow (2004)
13. Gennis, R.B.: Biomembranes Molecular Structure and Function. Springer, Heidelberg (1989)

Simulation Technologies Supporting Collaborative Training for Emergency Medical Services Personnel

Ilona Heldal[1]([X]) and Lars Lundberg[2]

[1] Western Norway University of Applied Sciences, Bergen, Norway
ilona.heldal@hvl.no
[2] Centre for Prehospital Research, University of Borås, Borås, Sweden
lars.lundberg@hb.se

Abstract. Training collaboration and work in various contexts in the health sector may be supported by different technologies. Based on a practical study involving Swedish emergency medical services personnel and different simulator technologies, this paper exemplifies the benefit of using processes, narratives, simulations and serious games to design contextual training scenarios for activities that are not currently utilized. The aim is to illustrate resources needed, in terms of technologies, methods, expertise and time for planning and performing high-value contextual training. The results argue for the benefits of using technologies aligned by processes and applications defined by using narratives and serious games. While processes help to define a more efficient and effective training methodology; narratives and serious games are beneficial for designing scenarios to increase the learners experience. This knowledge is necessary for planning simulation rich scenarios for prehospital training.

1 Introduction

The health sector is undergoing a transformation, where the patients, the context of activities and collaboration between the involved organizations are attracting increased focus. This transformation requires increased changes in education and training for all professionals involved. Collaboration is difficult and 'offering effective education and training opportunities is a challenge in the absence of ubiquitous support, incentives, or [detailed and specific enough] requirements among health care professions' [1]. To understand new needs for education and training has societal importance.

Today much focus and resources are directed towards developing supportive, technology rich environments and centers. However, these environments are not easy to develop, and stakeholders are struggling with matching the new needs to existing state-of-the-art technologies and to training routines [2, 3]. There are several training environments where new, state-of-the art technologies e.g. the most modern virtual patients, videoconference solutions, virtual reality applications or complex simulators are available, but underutilized [4–6].

The aim of this paper is to illustrate resources needed for planning and performing simulation rich training scenarios for prehospital training. Discussing limitations and

© Springer International Publishing AG 2018
Y.-W. Chen et al. (eds.), *Innovation in Medicine and Healthcare 2017*, Smart Innovation,
Systems and Technologies 71, DOI 10.1007/978-3-319-59397-5_12

benefits of using the different technologies in empirical setting will inform the development and usage of simulations to address contextually relevant settings.

The focus is on EMS personnel, but lessons from other types of emergency management personnel, e.g. firefighters and police have been taken into consideration. This paper will not discuss communication technologies involving patients and relatives (e.g. emails, portals, mobile solutions, patient records).

2 Methodological Considerations

This is a qualitative study that examines the added value, resources and requirements of a prehospital training center with a focus on simulation and serious games (SSGs). The knowledge has been derived during development of the center and from developing training supported by several different technologies and expertise. Resources required and lessons learned for the examined activity flows are examined separately from the perspective of researchers, professionals conducting training, and learners.

To approach this development, a number of activities, decisions, changes, and thoughts about technology choices are described. This is more than a subjective journey, since activities to choose and use technologies supporting training were defined based on a previous literature review which considered results on efficient and effective support from technologies and high user experiences [3, 7]. In this paper, increased user experience refers to the sensory experience allowing to focus on application and to gain control on actions during these activity chains in an environment enriched with computer simulations. High experiences can be supported by using simulator technologies chosen for the corresponding activities such as for driving, for experiencing realistic patients' environment or for patient care [3]. One of the most difficult activities is to argue for advantages of one simulator technology over another. This is often context-specific, '*a given simulator may be more or less effective depending on the instructional objectives and educational context*'; therefore the focus on the mechanisms of instructional design features should not be neglected [4]. To set up technologies and to provide engaging context through these activity chains, methods and solutions from serious games were used. Data for this part comes from participatory observations and from relevant literature studies that identify beneficial trainings and settings.

Training was planned based on processes distilled with professional educators in order to define the basic and most necessary activity chains for the practice of contextual trauma training. The training scenario reported here, was created after interviewing EMS personnel and observing training situations and listening to descriptions of typical experiences encountered during prehospital practice informing technical experts about needs and possibilities to train [3, 8, 9]. The measures come from using a tool developed by distinguishing *immersiveness-triggers*, i.e. patterns indicating strong or weak involvement, and by counting these occurrences using video analysis [7].

3   Background Literature

To plan suitable settings and environments for prehospital training, one has to know the general influences that training need to take into consideration, existing beneficial training settings, and current limitations of using technologies. To overcome these limitations, possible solutions and basic concepts the actual study use is described.

Today's healthcare cannot bear its cost, is not robust, and does not satisfy a large number of patients [10]. Movement towards more efficient healthcare creates a requirement on improving healthcare education to prepare professionals to better utilize IT that may allow accessibility for health data and *'ensure inter-operability and promote technology and data exchange'* [11]. Most prehospital patient's medical care begins with a call to a dispatch center, a functional prehospital system is imperative for good healthcare [12].

There are a large number of success stories regarding the use of technologies in medicine. The following simulation technologies are most frequently used for prehospital training today: virtual patient simulators [13] have been used in of training and learning for more than 50 years (1959); driving simulators [14], allowing possibilities to learn how to drive in non-optimal, e.g. stressful situations [15]; experiences with scenario simulators, today, mainly developed for large crises [16]. In general, using simulator technologies and serious games is considered to contribute to better learning and risk perception [17]. There are several recent studies that concluded that technologies allowing realism and repeated *non-stressful* situations contribute to increased experience, which in turn can improve patient care [18], promise a better support for debriefing [19].

Another type of simulation relates to supporting increased understanding for activity flows by using processes. Processes are used to support planning actors, activities and other resources needed in order to achieve clear goals. The concept of processes is inherited from the manufacturing industry and has been used to assist different types of changes in the healthcare sector [20]. However, in many process descriptions the prehospital care is often treated as a single step in the systematic care taking process [20].

The large number of research results can be almost impossible to understand and apply for practitioners. Especially, since they can meet negative effects of using simulators [21], mainly short-term trials in different other contexts, qualitative results and single technologies. Setting up new technologies for training can be experienced as a deterrent factor on at least three different levels: (1) by experiencing insecurity when trying out the new technology together with the existing ones, (2) uncertainty about choosing the most suitable equipment, and (3) lack of confidence in using it properly and having it updated with proper features [22]. Benchmarks are difficult to define when results often are context dependent.

While there are needs to develop centers with modern technologies and methods to train different practitioners together [2, 3], there are no guidelines how to populate these with right technologies and activities. To assess and train main activities, especially for complex emergency situations involving professionals from the ambulance, rescue service and police is important. In order to collaborate easily with each other and be aware of the others' routines to follow similar training is beneficial [23]. Alfes and

Reimer identified as much as eight, out of thirty, missing activities when team leaders from the ambulance were training communicational handoffs with the fire rescue services. In general, in ten simulation exercises between the ambulance and the rescue service more than 50% of influencing handoff activities were missing [24]. By defining processes that influence training one can easier assess important parts influencing overall activities.

4   Practical Needs in Prehospital Training

Emergency management stakeholders from the Västra Götaland Regional Council (VGR), one of twenty regions/county councils in Sweden, decided to promote the use of simulation technologies in prehospital training. They wished to provide a better contextual understanding of everyday activity chains for professionals from the ambulance services. These chains usually begin with a call from the dispatch center and driving to the patient. After arriving at the scene, initial patient assessment is performed and, if needed, the patient is transported by ambulance to the hospital. Ongoing treatment is made en route and finally, the patient is handed over at the hospital [25]. Ideally, the training ends with a debriefing, providing a possibility to learn from experiences. The stakeholders realized the need to develop the actual prehospital training, and the possibilities of using simulation technologies for achieving this. They had already experienced a positive impact from training with patient simulators. Their vision was to build a center for education and training of certified professionals engaged in prehospital emergency care and to make collaboration easier between the different regions and other actors involved in emergency management. They experienced difficulties in: choosing the most suitable equipment, trying out possible solutions, and lacking confidence in adjusting new technologies for their own goals, they connected researchers interested in information technologies from the University of Skövde, and in prehospital education from the University of Borås.

5   Towards Collaborative Contextual Training

This section describes main work sequences towards developing a simulation-rich training center, where collaboration and contextual training for emergency management is possible. While the very first idea of such a place is older, this paper focuses on work towards this center completed in the last six years.

5.1   Preliminary Work

The preliminary work (2012–2013) focused on determining the basics: which are the best training centers in the area, what is done and how does the state-of-the-art from related research fields influence this study. The work resulted in several sub-group collaborations. A group of researchers and practitioners visited recognized, successful centers, e.g. the Institut. für Notfallmedizin und Medizinmanagement in Munich, to gain information on training procedures and technologies. Another group investigated processes influencing based on state-of-the-art models based on literature reviews, and yet another, followed and documented actual training. Actual training in this region of the country often involves the use of patient simulators or moulage patients. Sometimes only the most necessary objects required for patient treatments, or a classroom with a whiteboard were used for training.

Before defining the demonstration, three different groups of at least two researchers observed training sessions for ambulance crews in actual conditions. The scenarios were divided into five important activities from (1) Call activation, (2) driving to the emergency scene (the ambulance EMS personnel had to practice driving an ambulance car in the city) to access the patient, (3) examining the patient and making clinical decisions, (4) driving to the hospital, and (5) handing over the patient. Debriefing sections followed all main activities. The instructors (and the external observers) used observation sheets with focused items throughout. Training days also included lectures, for example interpreting ECGs. The ambulance EMS personal use to practice in small groups, consisting of two or three persons. It is worth noting that the different steps were decoupled and trained separately; hence the intersections and interactions between activities and organizations were not trained.

Already in this phase it became clear that some main activities from the whole activity chain in prehospital work were not, or very seldom, trained. Examples are finding/recognizing places where the emergency call was sent from, handling the patients' environment, driving and providing care for the patients during the transport to the hospital. According to our knowledge, the whole activity chain is rarely exercised in realistic settings. These aspects are important, especially finding/recognizing the emergency scene, and taking en route care of the patient in the ambulance. These missing training activities may be hindering time critical interventions.

During this step, descriptions from stakeholders and EMS personnel regarding actual work situations, used technologies, and experiences were recorded. A narrative here is to be understood as a story of the connected activities that are always completed during an emergency call activation. Scenarios examined included unusual experiences e.g. finding the correct entrance, dealing with aggressive pets, meeting or even handling untidy environments, can be considered as often recurring, but not necessarily trained for. These stories are not necessarily optimal, but follow a logical line familiar for the practitioners. This line can be followed to understand different training situations.

5.2 A Demonstration

The general purpose of the demonstration (performed 2013) included: (1) allowing EMS personnel to train the whole chain of activities for two separate scenarios that are usually included in their training, from getting the emergency activation to ending the activities, usually by patient handover at the emergency room, (2) using simulation technologies to maintain continuity and enhance realism, promoting naturalism in the flow of activities needed, and (3) allowing the performance of task and record activities.

The chosen scenarios were: (a) a gas station attendee called for an ambulance after a truck driver complained of chest pains, and (b) an ambulance is called for after a collision between a biker and a car. The used equipment used were: a patient simulator

SimMan 3G, a real hospital bed, the driving simulator from the university used for dispatching and driving through realistic geographical environment, a standard equipped ambulance (VW T5 Profile), a real truck cab (used here as exterior props and observation and communication place), a real bicycle, projectors projecting realistic environment around the accident, computers, an audio system allowing audio simulation for the road environment in the road traffic accident case, a 3 × 2 feet large projection surface projecting the scene of accident, miscellaneous equipment consisting of chairs and tables that were used as simple props, and three wireless cameras for recording the trials. The average total time for the whole scenario was approximately 29 min, divided in the following main activities: Preparatory activities (4 min); Driving to Emergency Scene (6 min); Clinical care delivery and loading the 'patient' (7.5 min); Transport to the cardiac ICU (5.5 min); Structured patient handover (6 min). Video clips presenting the actual scenarios are available on the internet [25] and some data about the usefulness of these simulations have been presented earlier [3].

Planning the study and choosing the technologies required several meetings. Performing the training with the EMS personnel involved experts from the informatics research group (eight people), from the prehospital training center (three instructors), two EMS personnel performed the training, and an ambulance doctor and an ED nurse who acted in the handover of the patients.

5.3 The Pilot Study

Since the value of using combinations of different simulator techniques had been demonstrated [3], the study continued with illustrating these benefits in a pilot study involving a large number of EMS personnel. Another lesson from the demonstration was the value of collecting evidences for debriefing and assessment, and for this an application (based on the process and including videos and other recording from the activities) was developed.

The pilot study lasted three days (November, 2014) and focused on the training of 24 EMS personnel from regional ambulance stations. The EMS personnel were exposed to a traditional *basic scenario* (i.e. the way they usually train) and a simulation-rich, high-end *contextualized scenario*, for two common medical conditions. Data was collected via recordings, questionnaires, interviews and observation lists. These were tagged and ordered so that specific procedures could be easily followed. The configuration of technologies was defined based on the demonstration and the process model including the main activities described: (1) Activation by dispatch center, (2) Driving to the patient, (3) Finding the place, (4) Assessment and Clinical Care delivery, (5) Transport to the hospital, (6) Patient handover, and (7) Debriefing.

Based on lessons from the demonstration, a simulation environment was established and a test schedule for 24 EMS personnel was developed. Almost all technologies from the demonstration were adjusted based on the process description for training, and two situations enriched with the narratives described by the practitioners from the pre-study and demo were developed. One of the scenarios used a common, everyday environment supported only by the patient simulator, and the other scenario (a rich scenario) followed the whole process, from the initial alarm call to the final debriefing at the hospital. The results argue for overall increased involvement in high-end settings and contributed to a discussion on the value of processes and strategies for such training [26]. A special instrument (ISRI, the Immersion Score Rating Instrument) was developed to estimate the level of immersion based on activities influencing involvement, and reported higher immersion values for the simulation rich settings [7].

This pilot study resulted in a number of questions that remains to be answered, for example the significance of using this planning for other scenarios, if technologies were adequate or set-up adequately for the goals, or if this type of training can be generalized to other training situations.

6   Discussion: Effectiveness, Efficiency and User Experiences

In order to discuss resources needed (see an overview in Table 1) for planning and performing simulation rich training scenarios for prehospital training at least three separate activity flows have to be taken in consideration:

- Planning overall activities to find evidences needed to argue for a training center to identify general resources needed to start a center.
- Distinguishing necessary training activities to start with – also a minimal amount of activities.
- Demonstrating which technologies are needed and useful.

Table 1. Technologies used in the pilot study

Main activities	Associated activities	Simulations	Other technologies
1 - Emergency activation	Receiving the alarm call from the dispatch center and approaching the car.	–	Communication devices
2 - Driving to Emergency Scene	Discussing and preparing for the case. Ongoing communication with the dispatch center and, if needed, to the receiving hospital.	Car simulator, scenario simulation	Loggings, instruments, communication devices
3 - Locating the Emergency scene	Using maps, communication	Scenario simulation	Communication devices
4 - Clinical care delivery	Taking care of the patient, the patient's place, preparing instruments, communicating with the dispatch center.	Patient simulator, scenario simulation	Instruments, communication devices, loggings, projections
5 - Transport to the ED	A complex activity, where the EMS personnel have to provide care, communicate, and prepare a report.	Car simulator, scenario simulation, patient simulator	Instruments, loggings, communication devices
6 - Structured Patient handover	Reporting, provoding care, communication.	Patient simulator	–
7 - Debriefing	Discussing symptoms, signs and activities.	–	Debriefing support

6.1   Towards Developing a Training Center

While there are motivations for more training and new, collaborative training for using simulation technologies (e.g. [8, 23]), more information is needed, with economic evidences, provision of increased training time, especially necessary when realistic training is not possible, or when large classes needs to be trained.

Besides operating costs for a training center, questions regarding ownership, accessibility and overall responsibility for locations may arise. Deciding the location for the center and planning the rooms and responsibilities resulted in a three-year delay (at the present stage) in starting the actual simulator center. Before using technologies there are several knowledge barriers to overcome: (1) the technical barriers – building up competence using the actual technologies, (2) the financial barriers – via a model that explains how value can be created and how the stakeholders can be satisfied, (3) the organizational barriers, in order that new technology fits in with the structures and processes in the adopting organizations, and (4) the behavioral barriers, so that co-workers in the organization are convinced about adopting the technology, e.g. by managing technology resistance, and inter-organizational barriers [27]. Several of these barriers were evidenced by the authors experience, who encountered questions such as; how the patient simulation works together with video or logging technologies? (technical barriers), which simulators are needed and handled? (financial barriers), who can be responsible for training and how? (organizational barriers). We met these barriers during the time the study was planned. Behavioral barriers were met during the pilot study, when EMS personnel compared e.g. the simulator with 'better situations', e.g. role playing or settings with more 'real settings' e.g. using their own ambulance bags. Using even more technologies, we also observed and managed interoperability issues between different simulation technologies, and differences in the EMS personnel operating guidelines personnel were familiar with.

6.2   Designing Training Activities at a Simulator Center

One of the aims of the present study is to understand if actual training activities can be performed at a larger simulator center, and what can be changed, or further developed. The training behind this research is based on contextualizing already trained clinical activities (see Table 1). While the actual clinical care activity was criticized in the pilot study, training other activities (especially planning for driving out to the patient, and later driving the patient back to the hospital) were considered as informative and useful experiences, as observed and later expressed in the interviews.

The necessary activities may also be completed with some unexpected situations from real narratives from experienced EMS personnel. These can be simulated and make the training a little bit different from situation to situation. By interviewing experienced EMS personnel, we learned how important it is be able to handle pets around patients, as well as to notice subtle information regarding orderliness and the occurrence of certain objects, arrangements and smells in the patient's home. Thinking about what is happening when the EMS personal drives to the destination defined easy-to-use situations attracting the attention of the subjects to the flow. For example, a large exit portal identical with the exit at the ambulance center were simulated and opened as they drove from the center. Almost all expressed their surprise during this opening and during driving out from usual streets. One EMS personnel exclaimed: '*I bought fish here yesterday, it was really good for dinner*' – when they passed the nearby grocery store.

6.3   Choosing Right Technologies at a Simulator Center

Practitioners do not necessarily know about the existence of many technologies, or to choose and use them on an everyday basis. Even though these technologies are available, additional information is needed to be able to use them on an everyday basis. For example, clinical decision support systems for prehospital care do exist, but there is little or no demand to implement those systems in the ambulance organizations. Debriefing and evaluation are yet not well supported by technologies; most of the evaluations are done manually. There are tables and web based forms, but it is difficult for educators and students to examine the large amount of data.

According to this study it would also be valuable to differentiate technologies needed to support training and examine separately: (1) technologies in everyday work in the prehospital care (e.g. new communication technologies and reporting devices), (2) technologies supporting recording and evaluations during training (e.g. video technologies or other logging systems), and (3) high-end technologies providing naturalistic environments (e.g. driving simulators and patient simulators). These have different goals and their usage needs different competences.

Setting up technologies for the pilot study demonstrated the possibility and the benefits of defining contextual training for EMS personnel. The process-based training approach showed several benefits, especially designing the flow needed during the activities in the environment. Defining the necessary activities in an activity chain can be simulated making the scenarios more believable. To use only one of the main technologies, the patient simulator, for a more complex activity is difficult for the trainers. In order to keep up with technology development, and to choose and argue for different technologies may be difficult in the current climate, if the center does not provide both contextual technologies for EMS training needs, as well as other technologies to support the training of professionals.

7   Conclusions

The main themes expressed in this paper are the illustration of the possibility and the benefits of narratives, simulation and of using processes to plan, set up and assess activities in a systematic way. The authors have indicated the importance to consider a larger context in training activity chains in the emergency care. As an example, the collaboration between EMS personnel in the ambulance, the patient, and people contributing with information from a dispatch center on the way to hospital is not trained. These activities should be trained, since they are of vital importance. While this work exemplified the benefit of varying situations, it also pinpoints the need for defining additional methodologies to understand the role of a basic, minimalistic scenario and how this can be supported or extended by technologies in systematic evaluations. A first step here is not only to provide seamless use of technologies, but also to support their seamless integration into education. Defining process description for training requires a clear understanding of the context with focus on necessary actors and technologies, and activity flows. This will result in clearly describing requirements, responsibilities and priorities, which may contribute more effective management in an interorganizational context. This is necessary since emergency management often includes different units.

Acknowledgement. Advanced paramedic Ricky Ellis (Dublin, Ireland) proofread the manuscript and provided valuable suggestions.

References

1. Walsh, L., et al.: Building health care system capacity to respond to disasters: Successes and challenges of disaster preparedness health care coalitions. Prehospital Disaster Med. **30**(02), 112–122 (2015)
2. Pellegrino, L., Reed, L.: Medical simulation centers: Selection and tips from the field. In: 2013 International Symposium on Human Factors and Ergonomics in Health Care: Advancing the Cause, Baltimore, MD (2013)
3. Backlund, P., et al.: Collaboration patterns in mixed reality environments for a new emergency training center. In: Proceedings of the IEEE European Modelling Symposium. IEEE, Manchester (2013)
4. Cook, D.A., et al.: Comparative effectiveness of instructional design features in simulation-based education: Systematic review and meta-analysis. Med. Teach. **35**(1), e867–e898 (2013)
5. Ellaway, R.H., Davies, D.: Design for learning: Deconstructing virtual patient activities. Med. Teach. **33**(4), 303–310 (2011)
6. Vuk, J., et al.: Impact of simulation training on self-efficacy of outpatient health care providers to use electronic health records. Int. J. Med. Inform. **84**(6), 423–429 (2015)
7. Hagiwara, M.A., et al.: Measuring participants' immersion in healthcare simulation: The development of an instrument. In: Advances in Simulation, vol. 1(1) (2016)
8. Schaafstal, A.M., Johnston, J.H., Oser, R.L.: Training teams for emergency management. Comput. Hum. Behav. **17**(5–6), 615–626 (2001)
9. Söderström, E., et al.: Combining work process models to identify training needs in the prehospital care process. In: Johansson, B., Andersson, B., Holmberg, N. (eds.) Perspectives in Business Informatics Research. Springer, Cham (2014)
10. Porter, M.E., Teisberg, E.O.: Redefining Health Care: Creating Value-Based Competition on Results. Harvard Business School Press, Boston (2006)
11. Graham, S.L., et al.: Report to the President and Congress Ensuring Leadership in Federally Funded Research and Development in Information Technology (NITRD) Program, pp. 1–68 (2015)
12. Bigham, B.L., et al.: Expanding paramedic scope of practice in the community: A systematic review of the literature. Prehospital Emerg. Care **17**(3), 361–372 (2013)
13. McFetrich, J.: A structured literature review on the use of high fidelity patient simulators for teaching in emergency medicine. Emerg. Med. J. **23**(7), 509–511 (2006)
14. Milleville-Pennel, I., Charron, C.: Driving for real or on a fixed-base simulator: Is it so different? An explorative study. Presence Teleoperators Virtual Environ. **24**(1), 74–91 (2015)

15. Lenné, M.G., et al.: Detection of emergency vehicles: Driver responses to advance warning in a driving simulator. Hum. Factors J. Hum. Factors Ergon. Soc. **50**(1), 135–144 (2008)
16. Martí, J., et al.: I2Sim modelling and simulation framework for scenario development, training, and real-time decision support of multiple interdependent critical infrastructures during large emergencies. In: NATO (OTAN) MSG-060 Symposium on How is Modelling and Simulation Meeting the Defence Challenges out to (2008)
17. Okuda, Y., et al.: The utility of simulation in medical education: What is the evidence? Mt Sinai J. Med. J. Transl. Personalized Med. **76**(4), 330–343 (2009)
18. Ilgen, J.S., Sherbino, J., Cook, D.A.: Technology-enhanced simulation in emergency medicine: A systematic review and meta-analysis. Acad. Emerg. Med. **20**(2), 117–127 (2013)
19. Hagiwara, M.A., et al.: Decision support system in prehospital care: A randomized controlled simulation study. Am J Em. Med **31**, 145–153 (2013)
20. Åhlfeldt, R.-M., et al.: Supporting active patient and health care collaboration: A prototype for future health care information systems. Health Inf. J. **22**(4), 839–853 (2015)
21. Bertolini, G., Straumann, D.: Moving in a moving world: A review on vestibular motion sickness. Front. Neurol. **7**, 14 (2016)
22. Takemoto, K., Motoya, Y., Kimura, R.: A proposal for effective emergency training and exercise program to improve competence for disaster response. Disaster Res. **5**(2), 197–207 (2010)
23. Vyas, D., et al.: Prehospital care training in a rapidly developing economy: A multi-institutional study. J. Surg. Res. **203**(1), 22–27 (2016)
24. Alfes, C.M., Reimer, A.: Joint training simulation exercises: Missed elements in prehospital patient handoffs. Clin. Simul. Nurs. **12**(6), 215–218 (2016)
25. A Training Center for Emergency Medical Services. https://www.youtube.com/watch?v=WGCgEzNwXDU. Accessed 2 Apr 2017
26. Heldal, I., Lundberg, L., Hagiwara, M.: Technologies supporting longitudinal collaboration along patients' pathway: Planning training for prehospital care. In: ICICKM: Academic Conferences International Limited, Bangkok, November (2015)
27. Suneson, K., Heldal, I.: Knowledge barriers in launching new telecommunications for public safety. In: ICIKM: Academic Conferences International Limited, Bangkok, November 2010

An Indentation-Type Instrument for Measuring Soft Tissue Elasticity

Zhongkui Wang[✉], Kousuke Kadoma, and Shinichi Hirai

Department of Robotics, Ritsumeikan University, Kusatsu, Shiga 525-8577, Japan
wangzk2011@gmail.com

Abstract. An indentation-type instrument was presented in this study for measuring soft tissue elasticity and for further applications in finite element modeling and simulation of soft organs. The instrument consists of a stepping motor, an elongated probe, a sphere at the end of the probe, a load-cell fixed inside the probe, and a frame base. Indentation experiments were firstly performed on samples made of artificial muscle material and elasticity was estimated through curve fitting. The estimated elasticity was compared with the value measured by Instron material testing machine and good agreement was achieved. The instrument was then used to measure the elasticities of bovine and chicken livers and results suggested that the extended contact theory with two parameters is necessary to obtain a precise approximation of the liver behaviors. Finally, the device was applied to identify embedded hard objects within a soft medium for mimicking hard tumor inside soft tissue scenario.

Keywords: Tissue elasticity · Probe instrument · Parameter estimation

1 Introduction

Surgical simulation provides a practical way for surgical training and planning prior to the actual surgery. To this end, accurate computer model of human organs and tissues are demanded. This in turn requires accurate physical properties of biological organs and tissues. However, measuring such properties in vivo is a very challenging task due to the complex internal environment and limited space inside body, and also limited by the complex mechanical properties of biological organs and tissues. To take this challenge, in this study, an indentation-type instrument was proposed and validated using artificial muscle material and biological liver samples.

In literature, many methods have been proposed to characterize the mechanical properties of biological tissues. For example, 1D tension tests were performed on human liver sample to characterize the viscoelastic and failure properties [1]. Shear wave transducer was used to measure shear mechanical impedance and complex shear moduli of porcine muscle, liver and kidney [2]. A force sensitive wheeled probe was presented to roll over the surface of soft tissue and measure the stiffness of the tissue [3]. An aspiration device was proposed for in

© Springer International Publishing AG 2018
Y.-W. Chen et al. (eds.), *Innovation in Medicine and Healthcare 2017*, Smart Innovation, Systems and Technologies 71, DOI 10.1007/978-3-319-59397-5_13

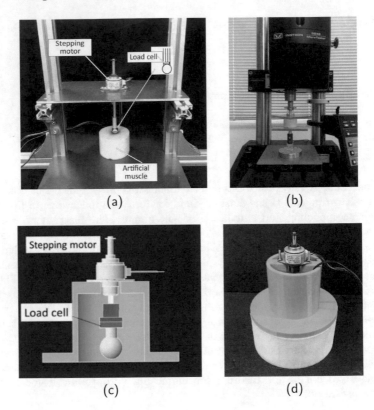

Fig. 1. The proposed prototype (a), the Instron material testing machine (b), the portable version of the instrument (c), and the application scenario (d).

vivo mechanical characterization of human liver [4]. Magnetic resonance (MR) elastography was also used to measure shear moduli of human liver [5,6]. Unfortunately, the most fundamental mechanical parameter, Young's modulus E, has not been addressed frequently in literature. Young's modulus has been widely used in finite element modeling and simulation of biological organs and tissues, and it dominates the tissue behaviors under the linear and small deformation assumption. For capturing nonlinear behaviors of biological tissues, the hyperelastic model has been widely used, but it usually involves more than three unknown parameters which make the parameter estimation difficult.

2   Method

2.1   The First Prototype

The prototype of the developed device (Fig. 1a) consists of a stepping motor to actuate the indentation motion, an elongated probe to transmit the motion, a sphere at the tip of the probe to perform the indentation, and a load-cell fixed inside the probe to measure the reaction force. Two kinds of samples were used in our experiments. The artificial muscle samples were made of silicon rubber materials, which reproduce the softness of human muscle. The second kind is biological liver samples. In order to obtain ground truth values of the artificial muscle elasticity, the Instron material testing machine (Fig. 1b) was employed in this study.

2.2   The Portable Device

To realize convenient measurement, a portable version was developed based on the prototype. A 3D printed frame was designed to hold and support the probe and the stepping motor (Fig. 1c). A flat circular surface was designed at the bottom of the instrument to press and stretch the measuring target to have a relatively flat contact area (Fig. 1d). The portable instrument can be easily held by a single hand of an adult.

2.3   Contact Mechanics

Most of human organs and tissues are supposed to be incompressible and the Poisson's ratio γ can be considered to be close to 0.5. In this study, we set Poisson's ratio γ as a constant of 0.49 and we were focusing on the measurement of linear elasticity, *i.e.* the Young's modulus E. According to Hertz contact mechanics, the contact force F between a sphere and a flat elastic surface is given by:

$$F = \frac{4}{3}\frac{E}{1-\gamma^2}R^{\frac{1}{2}}\delta^{\frac{3}{2}}, \tag{1}$$

where R denotes the radius of the sphere and δ denotes the indentation depth. The relationship was further extended to the following equation [7] to cope with more complex nonlinear behaviors by including another parameter B.

$$F = \frac{4}{3}\frac{E}{1-\gamma^2}R^{\frac{1}{2}}[\delta(1+B\delta)]^{\frac{3}{2}}. \tag{2}$$

2.4   Indentation Experiment and Curve Fitting

A 5 mm indentation trials was performed 5 times using the instrument and the indentation force was recorded by the load cell. Matlab curve fitting toolbox was used to fit the measured indentation-force curve based on the Eqs. 1 and 2. The artificial muscle samples have a size of 50 mm in diameter and 40 mm in height. Samples with two hardnesses (C0 and C5) were tested. To evaluate the results, same artificial muscle samples were tested on the Instron material testing machine.

Fresh bovine and chicken livers (Fig. 2) were obtained from a local butcher. They were cut into rectangular shape with sizes shown in Table 1. The bovine

(a)   (b)

Fig. 2. The bovine liver (a) and the chicken liver (b).

Table 1. The sample size of bovine and chicken livers

Liver	Length (mm)	Width (mm)	Thickness (mm)
Bovine	48	70	25
Chicken 1	105	46	19
Chicken 2	81	52	18
Chicken 3	72	66	14

sample was indented 5 times at 5 different locations to see the homogeneity. On the other hand, three chicken samples were prepared and each sample was indented only once. The indentation depth for both liver samples was approximately 3 mm.

3   Results

3.1   Property of Artificial Muscle

Figure 3 showed that both contact force expressions (Eqs. 1 and 2) can precisely approximate the experimental data. The best-fit parameters (average value of 5 trials) together with the ground truth data measured by Instron material test machine were listed in Table 2. The differences were insignificant among different methods. Since the artificial muscle samples have relatively linear properties, the results using Eq. 1 are quite close to the Instron test data.

3.2   Property of Biological Liver

Figure 4 showed the force-indentation curve of bovine and chicken livers. We did not find significant differences among different indentation locations for the bovine liver (Fig. 4a), which means the homogeneity of the liver is relatively good.

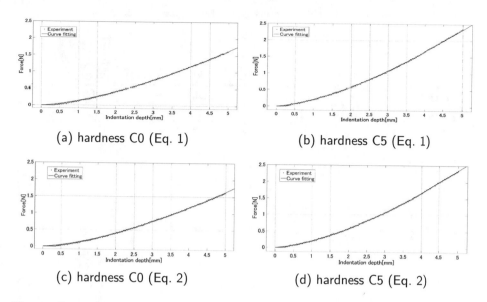

(a) hardness C0 (Eq. 1)   (b) hardness C5 (Eq. 1)

(c) hardness C0 (Eq. 2)   (d) hardness C5 (Eq. 2)

Fig. 3. Curve fitting results of two artificial muscle samples using different contact mechanics.

Table 2. Best-fit parameters of artificial muscle samples and the ground truth data from Instron test machine

Approximation method	Hardness C0		Hardness C5	
	E (kPa)	B (1/mm)	E (kPa)	B (1/mm)
Equation 1	38.0	—	53.9	—
Equation 2	37.5	0.0024	51.5	0.0078
Instron	38.2	—	53.2	—

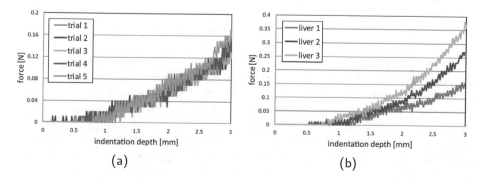

(a)   (b)

Fig. 4. The force-indentation relationships of the bovine (a) and chicken liver (b).

Table 3. Best-fit parameters of the bovine and chicken livers

Method	Bovine		Chicken 1		Chicken 2		Chicken 3	
	E (kPa)	B (1/mm)	E (kPa)	B (1/mm)	E (kPa)	B (1/mm)	E (kPa)	B (1/mm)
Equation 1	5.95	—	5.67	—	10.11	—	14.09	—
Equation 2	2.69	0.31	1.22	0.74	0.84	1.76	1.69	1.29

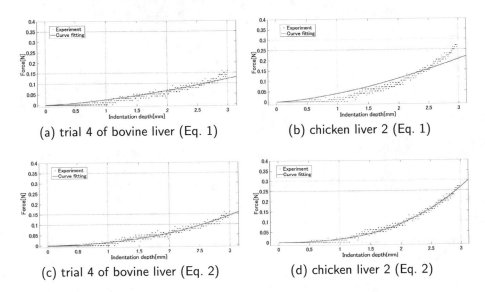

(a) trial 4 of bovine liver (Eq. 1) (b) chicken liver 2 (Eq. 1)

(c) trial 4 of bovine liver (Eq. 2) (d) chicken liver 2 (Eq. 2)

Fig. 5. Curve fitting examples of bovine and chicken livers using different contact force equations.

On the other hand, different chicken livers showed significant differences in terms of the force-indentation relationships (Fig. 4b).

By applying the proposed method, we could estimate the mechanical parameters of bovine and chicken livers. The results were listed in Table 3 and the curve-fitting examples were shown in Fig. 5. We found that the individual differences in chicken liver properties were quite large and they are relatively harder than the bovine liver. In addition, we found that the parameter E became smaller when using Eq. 2 to fit the force-indentation curve. The parameter B became significantly larger comparing with the artificial muscle results (Table 1). With using of Eq. 2, we could obtain much better curve-fitting results (Fig. 5). Apparently, the force-indentation relationships in biological liver cases have stronger nonlinearity, and an one-parameter model (Eq. 1) is no longer enough to precisely approximate the force-indentation curve. A model with at least two unknown parameters (Eq. 2) is necessary to ensure a better approximation. Accordingly, the original physical meaning of Young's modulus E was extended to the following expression $\hat{E}$ to include the nonlinearity with the increasing indentation depth [7].

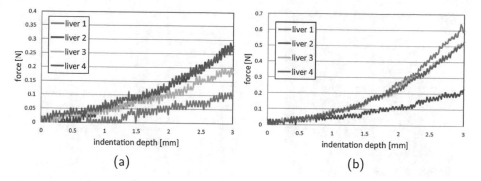

Fig. 6. Indentation results of the bovine liver (a) and chicken liver (b) using the portable device.

Table 4. Best-fit parameters of the bovine and chicken livers using the portable device and Eq. 2

Trial no.	Bovine liver		Chicken liver	
	E (kPa)	B (1/mm)	E (kPa)	B (1/mm)
Trail 1	2.89	0.164	11.53	0.312
Trial 2	8.44	0.135	11.20	−0.015
Trial 3	11.12	−0.035	14.96	0.147
Trial 4	13.57	−0.010	15.16	0.137

$$\hat{E} = E(1 + B\delta)^{\frac{3}{2}}. \tag{3}$$

3.3 Liver Test with the Portable Device

The portable device was also used to test the liver samples. To investigate the individual differences among different bovine livers, four bovine livers were tested. The indentation results were shown in Fig. 6. We also found significant individual differences in bovine livers as well as chicken livers. The Eq. 2 was used to fit the indentation results and the best-fit parameters were given in Table 4. We found that the differences were relatively large among different livers and the hardness of the liver was no longer determined by a single parameter E, but affected by both parameters of E and B. Integrating Eq. 2 into a finite element model will be one of our future works.

3.4 Identification of Embedded Object

One possible application of the developed portable device is to identify a hard object embedded inside a soft medium similar as a tumor embedded inside soft tissue. To this end, we constructed a phantom made of different artificial muscle

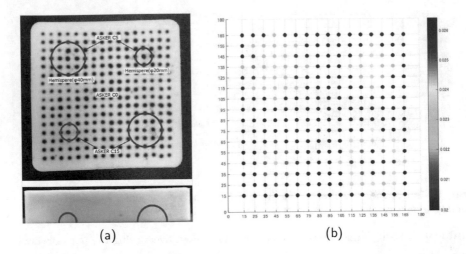

Fig. 7. A phantom made of different artificial muscle materials to mimic humor inside soft tissue: (a) the phantom from a top view (top) and a side view (bottom), and (b) the identified elasticity distribution.

materials (Fig. 7a). It consists of three materials of different hardnesses, which are the medium made of hardness C0 muscle, two hemispheres on the top side made of hardness C5, and two hemispheres at the bottom side made of hardness C15. The phantom has a size of 180 mm × 180 mm × 40 mm. The embedded depths were 20 mm and 30 mm for the large and small sphere respectively. Markers were drawn on the phantom surface to show the indentation locations. The developed device was used to indent over the phantom surface with a 5 mm indentation and generated an elasticity distribution as shown in Fig. 7b. We found that the device was able to identify an embedded sphere with a diameter of 20 mm and hardness is about two times of the surrounding soft tissue.

4   Conclusion

Measuring the mechanical properties of biological organs and tissues has always been an important topic in the field of computer-aided diagnosis and surgical training. Due to the nonlinearity of the biological tissues, it is hard to approximate with simple linear contact mechanics. In this study, we proposed a portable instrument for measuring soft tissue elasticity by applying extended Hertz contact mechanics. The instrument consists of a stepping motor to generate the indentation, a load cell to measure the contact force, and a probe with a sphere indentation head. The instrument was firstly used to measure the elasticity of two kinds of artificial muscle materials and results were validated by the Instron material test machine. Validation results showed that the traditional Hertz contact mechanics was good enough to capture the behavior of artificial muscle materials due to their linear properties. Secondly, the instrument was used to measure the elasticities of bovine and chicken livers. Significant individual differences were found in both livers. Traditional Hertz contact mechanics can no longer precisely approximate the nonlinear behaviors of the livers. The extended contact mechanics was able to fulfill this task, and the physical meaning of Young's modulus E was extended to a new parameter $\hat{E}$ which takes into consideration of indentation depth. A finite element model can be formulated using the new parameter $\hat{E}$. Thirdly, the portable instrument was used to identify the elasticity distribution of a soft artificial muscle phantom embedded with harder hemispheres to mimic the scenario of tumor inside soft tissue. Results showed that the proposed instrument can identify a diameter of 20 mm sphere with a hardness of two times larger than the surrounding soft tissue and embedded in a depth of 30 mm.

Acknowledgments. This work was supported by MEXT-Supported Program for the Strategic Research Foundation at Private Universities (2013-2017).

References

1. Kemper, A.R., Santago, A.C., Stitzel, J.D., Sparks, J.L., Duma, S.M.: Biomechanical response of human liver in tensile loading. Assoc. Adv. Automot. Med. **54**, 15–26 (2010)
2. Yang, X., Church, C.C.: A simple viscoelastic model for soft tissues in the frequency range 6–20 MHz. IEEE Trans. Ultrason. Ferroelectr. Freq. Control **53**(8), 1404–1411 (2006)
3. Noonan, D.P., Liu, H., Zweiri, Y.H., Althoefer, K.A., Seneviratne, L.D.: A dual-function wheeled probe for tissue viscoelastic property identification during minimally invasive surgery. In: IEEE International Conference on Robotics and Automation, pp. 2629–2634. Roma (2001)
4. Nava, A., Mazza, E., Furrer, M., Villiger, P., Reinhart, W.H.: In vivo mechanical characterization of human liver. Med. Image Anal. **12**, 203–216 (2008)
5. Asbach, P., Klatt, D., Hamhaber, U., Braun, J., Somasundaram, R., Hamm, B., Sack, I.: Assessment of liver viscoelasticity using multifrequency MR elastography. Magn. Reson. Med. **60**, 373–379 (2008)
6. Rouviere, O., Yin, M., Dresner, M.A., Rossman, P.J., Burgart, L.J., Fidler, J.L., Ehman, R.L.: MR elastography of the liver: preliminary results. Radiology **240**, 440–448 (2006)
7. Tani, M., Sakuma, A.: Applicability evaluation of young's modulus measurement using equivalent indentation strain in spherical indentation testing for soft materials. Jpn. Soc. Mech. Eng. **76**(761), 102–108 (2010)

A Collaborative and Mobile Platform for Medical Image Analysis: A Preliminary Study

Zhuofu Deng[1(✉)], Yen-wei Chen[2], Zhiliang Zhu[1], Youyin Wang[3], Yi Wang[2], and Min Xu[1]

[1] Northeastern University, Shenyang, China
dengzf@swc.neu.edu.cn
[2] Ritsumeikan University, Kyoto, Japan
[3] State Grid Liaoning Electric Power Company Limited Economic Research Institute, Shenyang 110015, Liaoning, China

Abstract. The tremendous advancement in mobile computing hardware and software changes everything in people including health care. Image review on computer-based workstations has made film-based review outdated. However, with the expansion of needs in portability of digital workstations, we wish a acceptable quality of medical image reviewing in a handled device, which could give radiologists more convenience. In this context, mobile health (m-Health) delivers health care services, overcoming geographical, temporal, and even organizational barriers. This paper presents a collaborative and mobile platform for image analysis which provides medical images shared so that the radiologists can review these data anytime and anywhere. In addition, this platform has the biggest differences with other popular or traditional systems that it supports collaboration and plug-in. It means that all the experts in the different locations can have a research together like in the same conference face to face and programmers can develop their favorite functions as they will. At last, the system also supplies many indispensable medical image analysis auxiliary functions such as image registration and image segmentation, both of which offer radiologists greatly exact diagnosis tools. We evaluated the platform with respect to expert satisfaction, functional characteristics. The experimental results show that our platform outperforms conventional image review systems in these regards.

1 Introduction

Digital radiography systems have been replacing traditional analog or screen-film systems over the last 3 decades [6]. The X-ray detectors can easily record the images in a computer-readable electronic format. This technologies have gained more and more popularities in many medical and industrial applications because they possess a number of useful capabilities such as electronic archiving, real-time acquisition and post image processing, etc. [5,14]. Nevertheless, these large equipments need constraints of space (i.e., consultation room) and time

© Springer International Publishing AG 2018
Y.-W. Chen et al. (eds.), *Innovation in Medicine and Healthcare 2017*, Smart Innovation, Systems and Technologies 71, DOI 10.1007/978-3-319-59397-5_14

(i.e., appointment) for taking doctor's diagnosis of medical image. Furthermore, since diagnostic medical images are made in various formats dependent on medical devices, their own methods to display and handle these diagnostic images are also additionally required [27]. For this reason, a portable device for medical image reviewing is needed.

The introduction of mobile device (PDAS) in the 90 s enabled physicians to easily download medical records, lab results, medical images, and drug information. Patients could be aware of their diagnostic disease control, and monitoring with comfortable mobile devices that accompany them everywhere [24]. In recent years, computing capability of mobile devices increases quickly and advanced compression algorithms are introduced to reduce the size of medical images [11,25], so that it is possible to show DICOM images on mobile devices now. So far, there are two categories of implementation for reviewing medical images in the smart cellphone, with the web technology [17,28] and the application in the mobile [9,15,18]. Although the platforms with web technology have great compatibilities so that it can be deployed into any operating system, unfortunately whether the browser or webview technique in Android or iOS give a bad interaction experience which is inappropriate for reviewing more complicated medical 2D or 3D images in the mobile. Therefore, it is more popular to achieve this with mobile application especially using multi-touches function [26].

2   Related Work

In general, there are two ways for reviewing medical images, local rendering and remote rendering. Local rendering means the medical data stored in the mobile will be rendered by its CPU and GPU for example OpenGL ES, and remote rendering refers that all of rendering work has been finished in the server, so the mobile just receive screenshots as result displayed on the screen [13]. In design of radiological mobile application, we always prefer remote rendering owing to the large quantity of medical image which cause the cost of computing time and battery energy consuming impossibly afforded by smart mobile devices. In remote rendering the mobile is only responsible for displaying surely only a little pressure. Moreover, with the rapid development of Internet and its infrastructure, the average speed of wireless network has increased at a rate similar to hardwired networks. Current devices on the fastest LTE broadband networks have speeds that range from 10 to 50 Mbps, compared with wired home Internet connection speeds in the range of 100 to 500 Mbps. Such wireless transmission speeds make mobile devices a more viable option for use in radiology and medicine [2]. In summary, the mobile is an excellent solution for solving the problems mentioned in the last section.

3D slicer and Pluto [19,22] are famous medical image system based on ITK [10] and VTK [23], both of which can not only support many familiar medical image processing functions like visualization, interaction etc., but also let radiologists have some functions like segmentation and registration work on it. It also permit developers programming the novel plug-in to satisfy their distinct demands and everyone can enjoy them. Despite the power they are limited only in PC or desktop consequently not good at portability anywhere and anytime. In [7], a non-real-time remote rendering system was introduced. The system generated 3D visualization medical images on the workstation in advance and then stored them in DICOM format. Mobile client can get the 3D visualization images as well as the ordinary DICOM images. As the 3D visualization images are not generated in real-time, the interaction of the client is very limited. Conversely, there are some platforms published in recent years with real-time that when requests come into server the rendering results return immediately [4, 16, 20, 21]. All the systems have a similar three layers architectures: one layer is the mobile display; the second layer is a proxy sever in charge of rendering and transmitting medical 2D or 3D image data; the last layer is a Picture Archiving System for retrieving medical digital files. These platforms permit users to have access to the database with mobile like iPhone and Android phone as they will and all difficulty and heavy rendering works have assigned to sever so client mobiles have little pressure. However, all of them are just reviewing ones where radiologists could not carry out some other common image processing work such as marking, registration and segmentation which is helpful for doctors to get more accurate diagnosis. Titinunt Kitrungrotsakul et al. [12] introduced a more sophisticated system in which except common remote rendering and reviewing mentioned before it can also allow clients have some advanced medical image processing operations in the mobiles connected with the 802.11g Wi-Fi network. In some case, for some divergent medical image data it needs some experts in different locations to have a group consultation, so collaboration can resolve this problem. Therefore, this system need to be optimized for better experience in this area. And if this platform also support plug-in, we believe it will be more popular specifically in mobiles.

In this paper, we address four challenges in constructing a system for collaborative and mobile platform of medical image analysis. At first, the radiologists can review the medical images on the smart phone or iPads. They can have a good interactions with the objects which is not affected by the network condition dramatically in real-time. The second challenge is that this platform provides clients have more complicated works on the 2D or 3D images convenient for their jobs. The third is supporting plug-in so that every programmer could develop new functions in server, as well as other clients can shared it at once in the mobile terminals. The last is the collaboration function. Multiple users who are in different place can work about the same image data jointly by using our system through their, smart terminals. For example, one student in place A is doing a organ segmentation using our segmentation plug-in function, while his teacher in different place B can not only see his segmentation result, but can also correct his segmentation result. This revolution will bring new and fantastic experience in the mobile radiology.

3   Methods

3.1   Platform Architecture

As depicted in Fig. 1, the proposed platform follows two-tier architecture, consisting of the mobile layer and a server layer. The server layer as the core of this platform is responsible for computing all the work occurred including rendering. In the server layer there are 8 modules for different jobs. We adopt VTK for displaying and interacting in server layer [8]. In user interface, we use QT as the panel development toolkit based on C++ which is popular today. Rendering computing is related to remote computing. For this, in a fixed time interval server will send the current screenshot image to clients in broadcasts for synchronization. Auxiliary tools contains marking, some sorts of segmentation and other elementary methods for processing medical image. Other modules will be introduced later. In the mobile client server there are also 8 modules and their functions are similar.

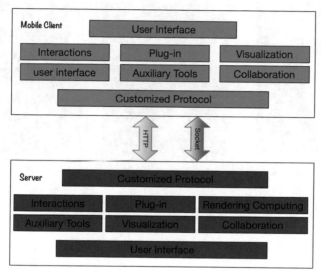

Fig. 1. Architecture of the proposed platform

3.2   Remote Rendering

Figure 2 presents the results of 2D and 3D medical images on the iPad Air 2 with remote rendering technology. As before mentioned, remote rendering depends on the sreenshot images produced by server in the network. Server local rendering is relied on VTK.VTK is an open-source, freely available software system for 3D computer graphics, image processing and visualization. It consists of a C++ class library and supports a wide variety of visualization algorithms including scalar, vector, tensor, texture and volumetric methods, as well as advanced modeling techniques such as implicit modeling, polygon reduction, mesh smoothing, cutting, contouring etc.

There are two forms of 3D reconstruction in image processing: surface rendering and volume rendering. Surface rendering simulates a three-dimensional data field with a series of slices, from which ROI (region of interesting) will be extracted and connected in triangular with some topologies. Its cost of computing time is lower and result 3D images are clear and interpreted. Nevertheless, because of only ROI, other related substance can not be observed at all so that it has no significance to radiologists at all. On the contrary, volume rendering will display everything in the 3D image which is helpful for the experts, although this method needs lots of CPU time and fine computer hardware configuration [3]. Because of the rapid advancement of software and hardware in computer science, our sever even a common PC can afford the plenty of volume rendering computing. So, in our platform we prefer volume rendering which benefits the radiology in server.

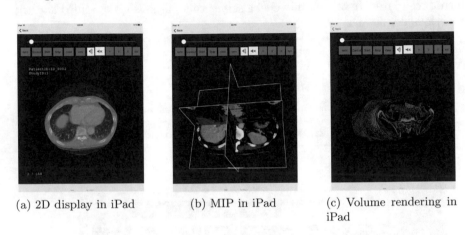

(a) 2D display in iPad   (b) MIP in iPad   (c) Volume rendering in iPad

Fig. 2. Remote rendering in iPad

3.3   Communicate Proctocol

In the platform, the custom communication protocol between the mobile client and the server is an application layer protocol based on Socket and Http as shown in Fig. 1. All the transmission files are XML file. The custom protocol is responsible for these tasks: screenshot images delivered from server and collaborative informations among clients. The first task focus on JPEG file transmission as JPEG is a welcome compression method not only higher speed but a small volume. In this task, we use socket protocol which is faster than HTTP and reduces our work on analysis of HTTP request. The second task guarantee perfect experience in medical teleconference with mobile terminals, that is sometimes when radiologists in different locations can have attended a common meeting together for the purpose of giving a more accurate diagnose with this proposed platform, for example they are seeking for synchronization of their voice, marks on the images, interactions with 3D images and so on. This custom protocol makes sure everything will happen. Accordingly, if we keep every clients' frame same, every user will see the same changes in the mobiles. So, this task will count the frames if there is a big latency in the Internet the mobiles will keep accordance. Http in our platform is responsible for transmitting some easy short parameters between two layers.

3.4   Collaboration Algorithm

The idea of collaboration has been more and more important in teleconference with computer techniques. It can make radiologists located in different places work together just like in the same meeting room: they can have a research for the difficult disease together; they can have chats with each other; if an expert rotates the object in 3D medical image wishing others to notice this portion of organism which has a little problem, the remaining experts will watch it almost in the same time. Our synchronized platform depicted in Fig. 3. Our collaboration function can let any client to be controllers whose manipulations should be seen in other iPads. So, we have an algorithm that sends the current controller's hand gesture to server for controlling. The collaboration algorithm is shown as Algorithm 1:

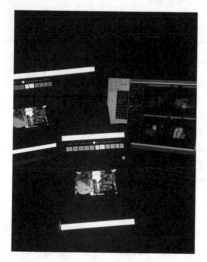

(a) Two iPads in marking

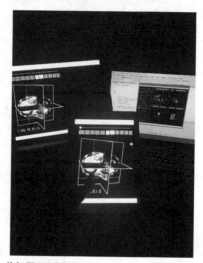

(b) Two MIP synchronized in two iPad

Fig. 3. Mobiles in collaboration

3.5   Plug-in

An excellent platform should be open source and be extended easily as your system that do not need other programmers' maintain. At this point, we integrated the idea of plug-in into our platform with an open source framework called Pluma [1] which is based on C++ and used for plug-in management. According to Pluma regulation, we design some complete and convenient standards where the users of platform can implement any plug-in if they like. Because all the projects are based on ITK and VTK, the plug-in is required in them also.

Algorithm 1. collaboration algorithm

System start:

1: server load image data
2: server screenshots 10 frames
3: **while** server is listening terminals' entries **do**
4:     **if** terminal has been linked with server **then**
5:         terminal sends its client name and adresss
6:         server send current user to this client
7:         server send frames to client 10 fps in broadcast
8:     **else**
9:         server send frames to client 10 fps in broadcast
10:     **if** some client set itself as current controller **then**
11:         the clients will send its hand gesture to server
12:         this transmitted gesture will control the server image objects
13:         server send frames to client 10 fps in broadcast
14:     **else**
15:         server send frames to client 10 fps in broadcast

What is more, these plug-ins are located in server not in mobiles. Therefore, we design a page of remote plug-in management in the smart terminals. Once you have an entry into server, the mobile system will search how many plug-ins in the server and will list all of them in the page of mobile system. Clients could select one of them to have a try in the device, and certainly all of the computing work will be finished by the server. Figure 4 presents the list of plug-ins we tested and the result of image processing result in the iPad.

4   Experiment and Result

This section discusses the performance of the proposed medical platform. In this study, we established two experiments for justifying the platform satisfy the needs of remote medical image analysis and can be deployed in any extremely complicated environment. Both of the two tests had been done under the same hardware conditions: the mainPACS server ran on a iMac with one 3.2 GHz Intel Core i5 CPU, 16 GB RAM and a AMD Radeon R9 M390 (2 GB) graphics card in the WLAN condition; mobile applications ran on iPad Air2.

In the first experiment, its purpose chiefly stressed evaluating the bandwidth occupied by the application in the mobile terminals under 4G network so that it could judge whether the proposed platform can run normally in the extreme situations. Figure 4 shows quantity of bandwidth in one day per hour the mobile devices need if there is a remote conference happening. In this diagram, we can find that with the time variation the bandwidth platform needs for normal running maintained between 83.1 Kb/s and 95.2 Kb/s almost a constant value. These minor changes are caused by other threads which had access to Internet in the iPad. This experiment result proved that the proposed platform would not need big bandwidth and will be affected by the Internet a little. In the second

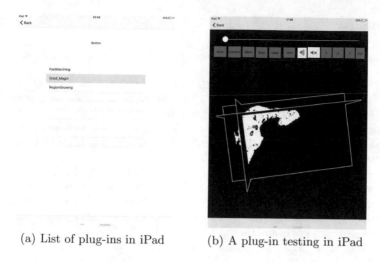

(a) List of plug-ins in iPad   (b) A plug-in testing in iPad

Fig. 4. Plug-in in mobiles

experiment, we had deployed this application on 19 medical cooperative partners' mobile devices in different places. During 9:30 am to 11:30 am and 7:00 pm to 8:00 pm when condition of Internet is worst in the working day, testing result of the proposed platform was accepted by all 19 clients who showed no inclination to feel the collaboration has been limited by crowed Internet.

5 Conclusion

A collaborative and mobile platform for medical image analysis is proposed in this paper. Classic medical image analysis systems such as 3D Slicer and Pluto are designed for single user and do not have collaborative and mobile functions. On the other hand, most of existing mobile medical systems are focused on medical visualization or medical education and not for medical image analysis. Our proposed system combines merits of the classic medical image analysis systems and existing mobile medical systems. Our system can used for joint works of multiple users in different places and provide a framework of plug-in for medical image analysis and visualization. Our future work will focus on the reduction of bandwidth and improvement of visualization and plug-in for real clinical applications.

Acknowledgments. This research was supported in part by the Grant-in Aid for Scientific Research from the Japanese Ministry for Education, Science, Culture and Sports (MEXT) under the Grant No. 15H01130 and No. 16H01436, in part by the MEXT Support Program for the Strategic Research Foundation at Private Universities (2013-2017).

References

1. Pluma: an open source C++ framework for plug-in management. http://pluma-framework.sourceforge.net
2. Auffermann, W.F., Chetlen, A.L., Sharma, A., Colucci, A.T., DeQuesada, I.M., Grajo, J.R., Kung, J.W., Loehfelm, T.W., Sherry, S.J.: Mobile computing for radiology. Acad. Radiol. **20**(12), 1495–1505 (2013)
3. Calhoun, P.S., Kuszyk, B.S., Heath, D.G., Carley, J.C., Fishman, E.K.: Three-dimensional volume rendering of spiral ct data: theory and method 1. Radiographics **19**(3), 745–764 (1999)
4. Celi, L.A., Sarmenta, L., Rotberg, J., Marcelo, A., Clifford, G.: Mobile care (moca) for remote diagnosis and screening. J. Health Inf. Dev. Countries **3**(1), 17 (2009)
5. Cho, H., Jeong, M., Han, B., Kim, S., Lee, B., Kim, H., Lee, S.: Development of a portable digital radiographic system based on FOP-coupled CMOS image sensor and its performance evaluation. IEEE Trans. Nuclear Sci. **52**(5), 1766–1772 (2005)
6. Cho, H.M., Kim, H.J., Lee, C.L., Nam, S., Jung, J.Y.: Imaging characteristics of the direct and mobile indirect digital radiographic systems. In: Nuclear Science Symposium Conference Record, NSS 2007, vol. 5, pp. 3840–3846. IEEE (2007)
7. Choudhri, A.F., Radvany, M.G.: Initial experience with a handheld device digital imaging and communications in medicine viewer: osirix mobile on the iphone. J. Digit. Imag. **24**(2), 184–189 (2011)
8. Hartkens, T., Rueckert, D., Schnabel, J.A., Hawkes, D.J., Hill, D.L.: VTK CISG registration toolkit an open source software package for affine and non-rigid registration of single-and multimodal 3d images. In: Meiler, M., Saupe, D., Kruggel, F., Handels, H., Lehmann, T.M. (eds.) Bildverarbeitung für die Medizin 2002, pp. 409–412. Springer, Heidelberg (2002)
9. Hsieh, J.C., Lo, H.C.: The clinical application of a PACS-dependent 12-lead ECG and image information system in E-medicine and telemedicine. J. Digit. Imag. **23**(4), 501–513 (2010)
10. Ibanez, L., Schroeder, W., Ng, L., Cates, J.: The ITK software guide (2005)
11. Ivetic, D., Dragan, D.: Medical image on the go!. J. Med. Syst. **35**(4), 499–516 (2011)
12. Kitrungrotsakul, T., Dong, C., Tateyama, T., Han, X.H., Chen, Y.W.: Interactive segmentation and visualization system for medical images on mobile devices. J. Adv. Simul. Sci. Eng. **2**(1), 96–107 (2015)
13. Lamberti, F., Sanna, A.: A streaming-based solution for remote visualization of 3d graphics on mobile devices. IEEE Trans. Vis. Comput. Graph. **13**(2), 247–260 (2007)
14. Lança, L., Silva, A.: Digital radiography detectors: a technical overview. In: Digital Imaging Systems for Plain Radiography, pp. 9–19. Springer (2013)
15. Lee, S., Lee, T., Jin, G., Hong, J.: An implementation of wireless medical image transmission system on mobile devices. J. Med. Syst. **32**(6), 471–480 (2008)
16. Lin, M.K., Nicolini, O., Waxenegger, H., Galloway, G., Ullmann, J., Janke, A.: Interpretation of medical imaging data with a mobile application: a mobile digital imaging processing environment. Front. Neurol. **4**, 85 (2013)
17. Lipton, P., Nagy, P., Sevinc, G.: Leveraging internet technologies with dicom wado. J. Digit. Imag. **25**(5), 646–652 (2012)
18. Mitchell, J.R., Sharma, P., Modi, J., Simpson, M., Thomas, M., Hill, M.D., Goyal, M.: A smartphone client-server teleradiology system for primary diagnosis of acute stroke. J. Med. Internet Res. **13**(2), e31 (2011)

19. Mori, K.: Navi-cad: navigation-based intelligent computer aided diagnosis system. Med. Imag. Tech. **24**, 173–180 (2006)
20. Nakata, N., Kandatsu, S., Suzuki, N., Fukuda, K.: Informatics in radiology (info RAD) mobile wireless DICOM server system and PDA with high-resolution display: feasibility of group work for radiologists 1. Radiographics **25**(1), 273–283 (2005)
21. Parikh, A., Mehta, N.: PACS on mobile devices. In: SPIE Medical Imaging, pp. 94180F–94180F. International Society for Optics and Photonics (2015)
22. Pieper, S., Halle, M., Kikinis, R.: 3d slicer. In: 2004 IEEE International Symposium on Biomedical Imaging: Nano to Macro, pp. 632–635. IEEE (2004)
23. Schroeder, W.J., Avila, L.S., Hoffman, W.: Visualizing with VTK: a tutorial. IEEE Comput. Graph. Appl. **20**(5), 20–27 (2000)
24. Silva, B.M., Rodrigues, J.J., de la Torre Díez, I., López-Coronado, M., Saleem, K.: Mobile-health: a review of current state in 2015. J. Biomed. Inf. **56**, 265–272 (2015)
25. Špelič, D., Žalik, B.: Lossless compression of threshold-segmented medical images. J. Med. Syst. **36**(4), 2349–2357 (2012)
26. Székely, A., Talanow, R., Bágyi, P.: Smartphones, tablets and mobile applications for radiology. Eur. J. Radiol. **82**(5), 829–836 (2013)
27. Treichel, T., Gessat, M., Prietzel, T., Burgert, O.: Dicom for implantations–overview and application. J. Digit. Imag. **25**(3), 352–358 (2012)
28. Valente, F., Viana-Ferreira, C., Costa, C., Oliveira, J.L.: A restful image gateway for multiple medical image repositories. IEEE Trans. Inf. Technol. Biomed. **16**(3), 356–364 (2012)

Detection of Liver Tumor Candidates from CT Images Using Deep Convolutional Neural Networks

Yoshihiro Todoroki[1], Xian-Hua Han[2], Yutaro Iwamoto[1], Lanfen Lin[3],
Hongjie Hu[4], and Yen-Wei Chen[1,3(✉)]

[1] Information Science and Engineering, Ritsumeikan University, Shiga, Japan
chen@is.ritsumei.ac.jp
[2] Faculty of Science, Yamaguchi University, Yamaguchi, Japan
[3] College of Computer Science and Technology, Zhejiang University, Hangzhou, China
[4] Radiology Department, Sir Run Run Shaw Hospital, Medical School, Zhejiang University,
Hangzhou, China

Abstract. There are multiple types of tumors occurring in the liver. Different tumors have different visual appearance and their visual appearance changes after injection of the contrast medium. So detection of liver tumors is considered as a challenging task. In this paper, we propose a method for detection of liver tumor candidates from CT images using a deep convolutional neural network. Experimental results show that we can significantly improve the detection accuracy by using our proposed method compared with the previous researches.

Keywords: Deep learning · Computer aided diagnosis (CAD) · CT image · Tumor candidate detection

1 Introduction

According to the latest statistics of population dynamics in Japan, the number of death according to the death cause ranking is higher in order of malignant neoplasia, heart disease, pneumonia, cerebrovascular disease. And the number of deaths due to malignant neoplasms is increasing year by year [1]. Early detection of lesions is important for effective treatments of the cancer. Diagnostic imaging is the most common method for finding cancers. Especially, recent remarkable progresses in medical imaging systems have enabled the acquisition of high-resolution CT or MRI datasets, so that by the use of high-resolution CT images, detection of small tumors (early stage) becomes possible. In image-based diagnosis, doctors judge the presence or absence of a tumor slice by slice based on their subjective medical interpretation, which is dependent on doctors' experience and is a time-consuming and labor-intensive task. Therefore, automatic tumor detection in CT images has become one of the most important research topics.

Many methods based on computer vision and image recognition techniques have been proposed for automatic tumor detections such as the K-means clustering method [2], the method based on first-order statistical features [3], the neural network [4], the Gaussian mixture model [5], the expectation maximization and maximization of the posterior marginal (EM/MPM) algorithm [6], sigmoid edge model [7]. In our previous work, we proposed a Bayesian Model for tumor detection [8]. On the other hand, there are multiple types of tumors occurring in the liver. Typical contrast-enhanced (multi-phase) CT images of five hepatic tumors (Cyst, focal nodular hyperplasia (FNH), cholangio cellular carcinoma (CCC), Hepato cellular carcinoma (HCC), metastasis (Meta)) are shown in Fig. 1. In this paper, we use the contrast-enhanced CT scans with three phases before and after the injection of contrast. A non-contrast enhanced (NC) scan is performed before contrast injection. After-injection phases include the arterial (ART) phase (30–40 s after contrast injection), portal venous (PV) phase (70–80 s after contrast injection). As shown in Fig. 1, we can see that different tumors have different visual appearance and their visual appearance changes after the contrast injection. So it is difficult to detect all types of tumors by one method. In this paper, we propose a method for detection of liver tumor candidates from CT images using a deep convolutional neural network.

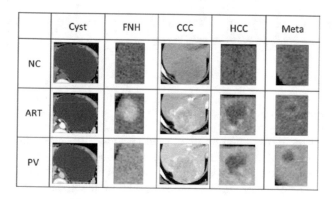

Fig. 1. Evolution patterns of five tumors over three phases.

The rest of the paper is organized as follows. In Sect. 2, we describe the proposed DCNN-based liver tumor detection. Experimental results are presented in Sect. 3. Section 4 provides our conclusion.

2 Proposed Method

The proposed method is based on deep learning. Deep learning is one of machine learning methods, which uses multilayered convolutional neural networks, and it exhibits high performance against various applications such as image recognition and language processing.

The proposed tumor detection method consists of two steps. The first step is to segment liver from the CT image using our developed liver segmentation algorithms [9, 10]. The second step is to calculate the probability of each pixel in the segmented liver belonging to tumors by the use of a deep convolutional neural network (DCNN) [11]. The DCNN we used is a network consisting of two convolution layers, two pooling layers and one full connection layer. The convolution layers are used to extract useful features. The pooling layers are used to minimize the spatial variations of the features and the last full connection layer is used for classification (calculation of the tumor probability). The DCNN for tumor detection is shown in Fig. 2.

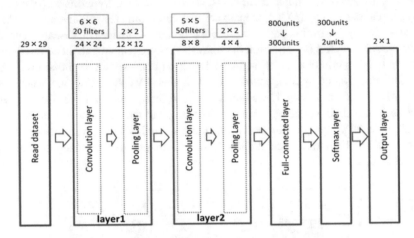

Fig. 2. DCNN for tumor detection.

2.1 Convolution Layer

The input is a patch with the target pixel at the center. The patch size is 29×29. In convolution layers, the feature extraction is done by convolving the input with filters. The filter sizes are 6×6 and 5×5 for the first and the second convolution layers, respectively. The numbers of the filers are 20 and 50 for the first and the second convolution layers. By denoting the r-th filter output of the l-th layer as O_r^l, and the k-th filter output of the previous layer as O_k^{l-1}.

$$O_r^l = \varphi\left(\sum_k W_{kr}^l * O_k^{(l-1)} + b_r^l \right) \tag{1}$$

where W_{kr}^l and b_r^l are filter and bias term. The * denotes the convolution operation and the φ is non-linear activation function. We used the same rectifier linear unit (RELU) function for all the neurons of the network except the top layer, the RELU function defined by

$$\varphi(x) = \max(0, x) \tag{2}$$

2.2 Pooling Layer

After the convolution layers, max-pooling layers are used to reduce the size of the feature by merging group of neurons. A max pooling layer shifts a 2×2 over the feature map and select only the highest value over each position of the pooling window as Eq. (3).

$$u_{ijk} = \max_{(p,q) \in P_{ij}} z_{pqk} \tag{3}$$

2.3   Learning and Output

The output of the network is the probability of the center pixel (target pixel) in the input patch. There are two output neurons in the network. One is the non-tumor probability (0~1) and another is the tumor probability (0~1). If the target pixel is labeled as tumor, the output will be (0, 1), while if the target pixel is labeled as non-tumor, the output will be (1, 0). A cross entropy function is used as a loss function to measure the accuracy of learning. We use this function to calculate the loss error and update the weights.

Suppose the number of training samples is N, the number of classes is K ($K = 2$), each training sample is represented by $\mathbf{x}$, its label is $\mathbf{t}$ and its network output is $\mathbf{y}$. The loss function for estimation of filter $\mathbf{w}$ is shown in Eq. (4).

$$E(\mathbf{w}) = - \sum_{n=1}^{N} \left(\sum_{m=1}^{K} t_{nm} \log y_{nm} \left(\mathbf{x}_n ; \mathbf{w} \right) \right) \tag{4}$$

3   Experiment

3.1   Dataset

3D multi-phase contrast-enhanced CT images of the liver from 75 cases are used in our research. The 75 cases consist of five types, namely, cyst, focal nodular hyperplasia (FNH), hepatocellular carcinoma (HCC), hemangioma (HEM) and metastasis (METS). Typical multi-phase images are shown in Fig. 1. Each type has 15 cases. Since each case has three phases (NC, ART and PV), we have 225 CT volumes. Among 75 cases, 50 cases (10 cases for each type) are used as training samples and 25 cases (5 cases for each type) are used as test. The liver was segmented from each CT volume by the use of our developed segmentation algorithms [9, 10]. Each pixel (or voxel) in the segmented livers was labeled by doctors with 0 (non-tumor pixel) or 1 (tumor pixel).

We randomly select 30,000 tumor pixels and 30,000 non-tumor (liver or vessel) pixels from training data set as training and validation samples. The number of training samples is 48,000 and the number of validation samples is 12,000.

3.2   Results

Typical detection results for different types are shown in Fig. 3. Figures at the top are original CT images with detected boundary by doctors. Figures at the bottom are detection results by our proposed method. The detection accuracies are summarized in Table 1. The denominator of each cell in Table 1 is the number of tumors (ground truth), which are detected by doctors, and the numerator is the number of correctly detected tumors. If the detected candidate has an overlap region with the ground truth, we define the detected results as correct answers. In order to make a comparison, we also show

144   Y. Todoroki et al.

detection results by the Bayesian model [8] in Table 2 as well as Table 1. It can be seen that most tumors could be detected. The detection accuracy of our proposed DCNN method outperformed that of the Bayesian Model. The bold face results in Table 1 are those that gave better detection results than the previous Bayesian model. Only the result of HCC (PV phase) is lower than the Bayesian model.

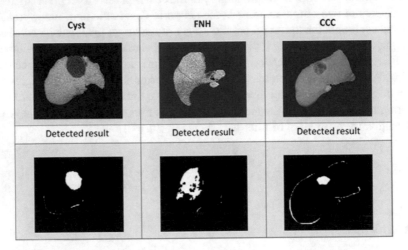

Cyst	FNH	CCC
Detected result	Detected result	Detected result

Fig. 3. Experiment results (top: CT images; bottom: detection results)

Table 1. Results of proposed method

	NC	ART	PV
Cyst	5/5	5/5	**5/5**
FNH	**3/5**	5/5	3/5
CCC	**5/5**	3/5	3/5
HCC	**3/5**	5/5	3/5
Meta	5/5	**5/5**	**5/5**

Table 2. Results of conventional method

	NC	ART	PV
Cyst	5/5	5/5	2/5
FNH	2/5	5/5	3/5
CCC	3/5	3/5	3/5
HCC	1/5	5/5	4/5
Meta	5/5	2/5	4/5

4   Conclusion

We proposed a robust liver tumor detection method using deep learning. Experimental results demonstrated that most tumors could be detected regardless of types and phases. The detection accuracy of our proposed DCNN method outperformed that of conventional methods.

Acknowledgments. This research was supported in part by the Grant-in Aid for Scientific Research from the Japanese Ministry for Education, Science, Culture and Sports (MEXT) under the Grant No. 15H01130, 15K00253 and No. 16H01436, in part by the MEXT Support Program for the Strategic Research Foundation at Private Universities (2013-2017), and in part by the Recruitment Program of Global Experts HAIOU Program from Zhejiang Province, China.

References

1. National Cancer Center, Japan: Center for Cancer Control and Information Services. http://ganjoho.jp/public/statistics/pub/statistics01.html
2. Ramaraju, P.V., et al.: Feature based detection of liver tumor using K-means clustering and classifying using probabilistic neural networks. Int. J. Eng. Comput. Sci. **4**, 11910–11915 (2015)
3. Ali, A.H., et al.: Diagnosis of liver tumor from CT images using first order statistical features. Int. J. Eng. Trends Technol. **20**, 155–158 (2015)
4. Mala, K., et al.: Neural network based texture analysis of liver tumor from computed tomography images. Int. J. Biomed. Sci. **2**, 33–40 (2006)
5. Park, S.-J., et al.: Automatic hepatic tumor segmentation using statistical optimal threshold. In: Computational Science – ICCS 2005, vol. 3514, pp. 934–940. Springer, Heidelberg (2005)
6. Masuda, Y., et al.: Liver tumor detection in CT images by adaptive contrast enhancement and the EM/MPM algorithm. In: Proceedings of IEEE International Conference on Image Processing (ICIP2013), pp. 1453–1456 (2011)
7. Foruzan, A.H., Chen, Y.-W.: Improved segmentation of low-contrast lesions using sigmoid edge model. Int. J. CARS **11**, 1267–1283 (2016)
8. Konno, Y., et al.: Bayesian model for liver tumor enhancement. In: Chen, Y.-W., et al. (eds.) Innovation in Medicine and Healthcare 2016, pp. 227–235. Springer (2016)
9. Dong, C., et al.: Simultaneous segmentation of multiple organs using random walks. J. Inf. Process. Soc. Japan **24**(2), 320–329 (2016)
10. Dong, C., et al.: Segmentation of liver and spleen based on computational anatomy models. Comput. Biol. Med. **67**, 146–160 (2015)
11. LeCun, Y., Bengio, Y.: Convolutional networks for images, speech, and time-series. In: Arbib, M.A. (ed.) The Handbook of Brain Theory and Neural Networks. MIT Press (1995)

Computerized Features for LI-RADS Based Computer-Aided Diagnosis of Liver Lesions

Mingzhong Chen[1], Lanfen Lin[1(✉)], Qingqing Chen[2],
Hongjie Hu[3(✉)], Qiaowei Zhang[3], Yingying Xu[1],
and Yen-Wei Chen[1,4]

[1] College of Computer Science and Technology,
Zhejiang University, Hangzhou, China
llf@zju.edu.cn
[2] School of Medicine, Zhejiang University, Hangzhou, China
[3] Sir Run Run Shaw Hospital, Zhejiang University, Hangzhou, China
hongjiehu@zju.edu.cn
[4] College of Information Science and Engineering,
Ritsumeikan University, Kyoto, Japan

Abstract. Liver Imaging Reporting Data System (LI-RADS) aims to standardize liver lesion imaging findings and diagnostic reports, and it is used as an accurate noninvasive diagnosis and staging method of hepatocellular carcinoma (HCC) nowadays. In this study, we proposed several computerized features for LI-RADS based computer-aided diagnosis of liver lesions. We used several popular machining learning approaches for computerized LI-RADS classification (benign and malignant classification) with our proposed features. The performance of each method was evaluated by using ROC curve and the best AUC score was 0.965 reached by the gradient boosting classifier.

Keywords: LI-RADS · Computerized feature · Machine learning · Malignancy · Classification

1 Introduction

Medical imaging technology has progressed immensely in these last few decades. The medical imaging-based examinations play an important role for accurate diagnosis. In order to standardize liver lesion imaging findings and diagnostic reports, Liver Imaging Reporting Data System (LI-RADS) [1] has been developed and widely used as an accurate noninvasive diagnosis and staging method of hepatocellular carcinoma (HCC), which is the most common type of liver cancer. The LI-RADS (v2014) has a 5-step hierarchical classification process as shown in Fig. 1. The LI-RADS classifies the observation into five main categories: LR-1 (definitely benign), LR-2 (probably benign), LR-3 (intermediate), LR-4 (probably malignant (HCC)), LR-5 (definitely malignant (HCC)). The detailed information about LI-RADS is available on the website of American College of Radiology (ACR) [2] and Mitchell's study [3]. LI-RADS defined several main features and ancillary features for categorization of hepatic lesions. The effectiveness of major features has been studied and validated by Ehman [4].

© Springer International Publishing AG 2018
Y.-W. Chen et al. (eds.), *Innovation in Medicine and Healthcare 2017*, Smart Innovation, Systems and Technologies 71, DOI 10.1007/978-3-319-59397-5_16

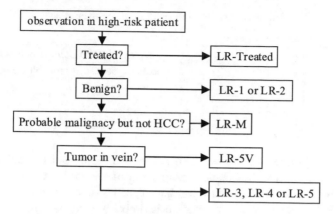

Fig. 1. LI-RADS classification process.

In spite of usefulness of LI-RADS, there are no computer-aided diagnosis systems (CADs) for liver lesions based on LI-RADS. Though there are some web applications developed for radiologists to follow the process and get the LI-RADS category [5], these applications require radiologists to extract or calculate features from the CT or MRI images manually. It is a time-consuming and labor-intensive task. So development of computable features based on LI-RADS is significant for radiologists. In addition, machine learning has been a popular method in computer-aid diagnostics. Shan [6] proposed a computer-aided diagnosis system for breast ultrasound using computerized BI-RADS (Breast Imaging Reporting and Data System) features and machine learning methods. In this paper, we proposed a set of computerized features based on medical definitions, LI-RADS descriptions and expert knowledge, which can be computed from CT or MRI images automatically. As a preliminary study, we also developed a simple computer-aided diagnosis (benign and malignant) system with our proposed computerized features and machine learning methods. In our CAD system, the observation is classified into two classes: benign and malignant, which will be an important information for doctors as a second opinion. By conducting the classification experiments, we can also validate the effectiveness of our proposed features and find a suitable machine learning method.

2   Computerized Features

2.1   Multi-phase CT Images

The database used in this study contains 53 liver Computed Tomography (CT) imaging cases. Each case contains three phases: Non Contrast (NC), Arterial (ART) and Portal Venous (PV). All of the cases were collected from Picture Archiving and Communication Systems (PACS) in Sir Run Run Shaw Hospital which were token in 2015. All of the cases were analyzed and reported by radiologists with five years' experience at least, and lesions with important findings were categorized (labeled) based on LI-RADS. Significant slices from each phase of each case were selected strictly by

Table 1. Detailed information of the liver cases

LR category	Number of cases	Description
LR-1	11	6 definite cysts and 5 definite hemangiomas
LR-2	11	5 probable cysts and 6 probable hemangiomas
LR-3	12	/
LR-4	9	/
LR-5, 5 V	10	/

radiologists for feature extractions. Each lesion was segmented by radiologists manually. In Table 1, we list the detailed information of the used liver cases. LR-Treated which means an observation of any category that has undergone loco-regional treatment and LR-M which means observation is probably malignant but not specific for HCC are not considered in this study for they do not indicate the malignancy of a liver lesion specific for HCC.

2.2   Computerized Features Based on LI-RADS and Expert Knowledge

Feature extraction is an important step in computer-aided diagnosis. In this paper, we defined and proposed a set of computerized features based on medical definitions, LI-RADS descriptions and expert knowledge. We defined features for both cases of benign and malignant.

2.2.1   Features Specified for Benign Cases

LR-1 and LR-2 are considered benign cases in LI-RADS. In medicine, there are many kinds of lesions which will be categorized into LR-1 or LR-2 [7]. We focus on two main benign lesions: cysts and hemangiomas. Features specified for them can make these lesions identified by models easily and lead to accurate estimations of malignancy. Other kinds of lesions can be added into our feature framework easily if necessary.

Cysts
Medical definition: A cyst is a fluid-filled enclosed cavity lined by benign epithelium. Typical multi-phase CT images of a definite cyst are shown in Fig. 2.

   LI-RADS description: round, smooth, liquid density, no enhancement

   Expert knowledge: clear boundary, no enhancement, round or round like, low density

Hemangiomas
Medical definition: A hemangioma is a common benign tumor consisting of vascular channels lined with endothelial cells. Typical multi-phase CT images of a definite hemangioma are shown in Fig. 3.

   LI-RADS description: peripheral discontinuous nodule like expanding enhancement or rapid enhancement, sharply circumscribed border, round like, oval or lobulated shape, low attenuation on unenhanced CT images

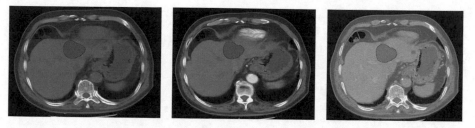

Fig. 2. Multi-phase CT images of a definite cyst (from left to right: NC, ART, PV).

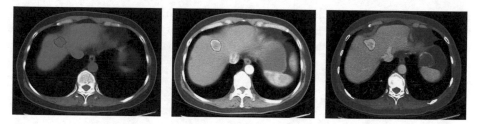

Fig. 3. Multi-phase CT images of a definite hemangioma (from left to right: NC, ART, PV)

Expert knowledge: round like, slightly lower density in NC phase, nodule like expanding enhancement.

According to the medical definition, LI-RADS description and expert knowledge of benign cases, we proposed and defined the features from four perspectives: shape, density, enhancement and boundary.

(1) **Shape features**

Region based roundness:

$$RBR_{PV} = eccentricity(oval(Lesion_{PV})) \tag{1}$$

Here, the *oval* function is to get an ellipse with the same second moment as the lesion, and the *eccentricity* function is to calculate the eccentricity of the given oval. It is a roundness measure from the region perspective.

(2) **Density features**

Average density:

$$AD_{NC} = \frac{\sum_i density(NC_i)}{area(Lesion_{NC})} \tag{2}$$

Here, the *area* function is to calculate the number of pixels where the lesion occupies. The summation function is to summarize the density of all the lesion points. This is an example feature from NC phase, and we also extract AD features from ART and PV phases, respectively.

(3) **Enhancement features**

Liver parenchyma based enhancement:

$$LPBE_{ART} = \frac{AD_{ART}}{avrDensity(LiverParenchyma_{ART})} \tag{3}$$

Here, the *avrDensity* function is to calculate the average density of the liver parenchyma, which is used to normalize the average density. This is an example feature from ART phase, and we also extract LPBE feature from PV phase, respectively.

Aorta based enhancement:

$$ABE_{ART} = \frac{AD_{ART}}{avrDensity(Aorta_{ART})} \tag{4}$$

Here, the *avrDensity* function is to calculate the average density of the aorta. It is an another normalization method which is based on the aorta density. This is an example feature from ART phase, and we also extract ABE feature from PV phase, respectively.

Hyper-enhancement percentage:

$$HEP_{ART} = \frac{\sum_i if(density(ART_i) > avrDensity(LiverParenchyma_{ART}))}{area(Lesion_{ART})} \tag{5}$$

Here, the *if* function returns 1 when the expression is true or 0 when the expression is false. The summation function is to get the number of pixels which satisfy the expression. It means the area percentage which is hyper-enhanced compared to the liver parenchyma. This is an example feature from ART phase, and we also extract HEP feature from PV phase, respectively.

(4) **Boundary features**

Boundary clearness:

$$BC_{PV} = \frac{\sum_i \left| \overline{out(PV_i)} - \overline{in(PV_i)} \right|}{length(boundary(Lesion_{PV}))} \tag{6}$$

Here, the *length* function is to calculate the number of pixels where the boundary of the lesion occupies. For each point on the border, we choose a certain number of points near the point and calculate the average density difference between the outside part and the inside part.

2.2.2 Features Specified for Malignant Cases

LR-4, LR-5, LR-5 V are considered malignant lesions specific for HCC. For LR-5 V is a special category of LR-5, we focus on the imaging features which can represent LR-4 and LR-5 and define the computerized features based on the LI-RADS definition, LI-RADS description and expert knowledge.

L4, L5

LI-RADS definition: high probability observation is HCC or 100% certainty observation is HCC. Typical multi-phase images of a definite HCC are shown in Fig. 4.

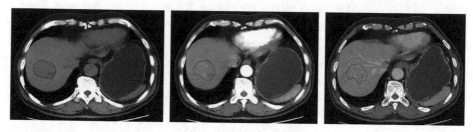

Fig. 4. Multi-phase CT images of a definite HCC (from left to right: NC, ART, PV)

LI-RADS description: arterial phase hyper-enhancement, wash out, capsule, diameter

Expert knowledge: indistinct boundary, uneven density, uneven enhanced, wash out

According to the LI-RADS definition, LI-RADS description and expert knowledge of malignant cases, we defined and proposed following features from the perspectives mentioned before.

(1) **Shape features**

Diameter:

$$D_{PV} = \max_{i,j} distance(PV_i, PV_j) \tag{7}$$

Here, the *distance* function is to calculate the Euclidean distance between two boundary points (*i* and *j*). It evaluates the size of the lesion.

(2) **Density features**

Boundary density:

$$BD_{ART} = \frac{\sum_i \overline{nearby(ART_i)}}{length(boundary(Lesion_{ART}))} \tag{8}$$

Here, the *length* function is to calculate the number of pixels where the boundary of the lesion occupies. For each point on the border, we choose a certain number of points near the point and calculate the average density, which is represented by the *nearby* function.

Capsule:

$$C = BD_{PV} \tag{9}$$

The capsule is a LI-RADS feature. It means a delayed enhancing rim. Lesion size (larger than 2 cm) and delayed enhancing rim are the strongest predictors of hepatocellular carcinoma [8]. The boundary density in PV phase can represent the capsule as a computerized feature.

(3) **Enhancement features**

Wash out:

$$WO = HEP_{ART} - HEP_{PV} \tag{10}$$

The wash out is a LI-RADS feature. It makes sense in the prediction of hepatocellular carcinoma [9]. It means visually assessed temporal reduction in enhancement relative to liver from ART to PV phase resulting in PV hypo-enhancement. When transforming the visual feature to computerized feature, we calculate the difference between the hyper-enhancement percentage of ART and PV phase. The percentage of ART is a benchmark and the difference shows how serious the hypo-enhancement is. HEP_{ART} and HEP_{PV} are hyper-enhancement percentages in ART and PV phase respectively, which are explained in Eq. (5).

From above, we deduced computerized features with detailed formulas, which will be used for machine learning based computer-aided diagnosis.

3   Computer-Aided Diagnosis Based on Machine Learning

In order to calculate the malignancy of a given lesion, we will consider the problem as a classification problem based on LI-RADS. In LI-RADS, there are some categories which indicate similar malignancy and too many levels are not friendly to the classification. So, level refactoring is necessary for the problem. For preliminary study, we will do 2-level refactoring according to the LI-RADS process and definition. Based on the level refactoring, we will calculate 2-level based malignancy. Now, we list the LI-RADS categories with definitions in Table 2.

Table 2. LI-RADS categories with definitions

LI-RADS category	Definition
LR-1	100% certainty observation is benign
LR-2	High probability observation is benign
LR-3	Both HCC and benign entity have moderate probability
LR-4	High probability observation is HCC
LR-5	100% certainty observation is HCC
LR-5 V	Observation is HCC invading vein

3.1 2-Level Based Malignancy Methods

In the 2-level based problem, we divide LI-RADS categories into two parts, one is the benign level and the other is the malignant level. It is a two-class classification problem obviously. According to the LI-RADS process and definition, judging whether a lesion should be categorized into LR-1 or LR-2 is an important step and also LR-1 and LR-2 are considered as benign. So, we divide LR-1 and LR-2 into the benign level and others into the malignant level.

3.1.1 Machine Learning Models
We used four popular machine learning models (decision tree, support vector machine, random forest, gradient boosting [10]) for our classification problem and compared the performance of them. All of the four models were implemented by python sklearn software package.

3.1.2 Performance Evaluation
For all the machine learning models, 5-fold cross-validation is carried out on the entire dataset to train and test the classifiers. For model evaluation, we use accuracy (ACC), precision (PRE), recall (REC), f1-score(F1), roc curve and area under curve (AUC) to evaluate the classification performance from different perspectives. The threshold is set as 0.5 for calculations of ACC, PRE, REC and F1.

3.1.3 Feature Selection
As we described in previous section, we proposed several computerized features to represent the imaging cases. In order to generate a good model, we use greedy strategy based feature selection to get a suitable feature set.

During the feature selection, we add one greedy feature into chosen features, which is chosen from the remaining feature pool and leads to a best performance promotion every iterative loop. Repeat the step until there is no feature can improve the performance. From all the evaluation scores, we choose AUC to evaluate the performance in each iterative loop.

3.2 Experimental Results

For choosing a better model, we contacted experiments using the four machine learning models and the greedy strategy based feature selection. When using Support Vector Machine Classifier, we selected the linear kernel for a better performance. All the iterative steps of feature selection with detailed scores are listed in Tables 3, 4, 5, and 6.

Table 3. 2-level based malignancy performance using gradient boosting classifier

Iteration	Greedy feature	AUC	ACC	PRE	REC	F1
1	LPBE (PV)	0.7915	0.6781	0.81	0.6428	0.6793
2	AD (PV)	0.9066	0.8290	0.8666	0.8666	0.8569
3	BC (PV)	0.9566	0.9036	0.9314	0.9	0.9118
4	WO	**0.965**	0.8654	0.8778	0.9	0.8831

Table 4. 2-level based malignancy performance using random forest classifier

Iteration	Greedy feature	AUC	ACC	PRE	REC	F1
1	AD (PV)	0.7220	0.6781	0.7898	0.6666	0.7005
2	BC (PV)	0.8696	0.7709	0.8214	0.7761	0.7917
3	HEP (PV)	0.8835	0.8290	0.8664	0.8333	0.8440
4	RBR	0.8895	0.8490	0.8476	0.9047	0.8739
5	C	**0.8911**	0.8109	0.8628	0.8047	0.8287

Table 5. 2-level based malignancy performance using decision tree classifier

Iteration	Greedy feature	AUC	ACC	PRE	REC	F1
1	BC (PV)	0.7364	0.7363	0.7795	0.7428	0.7521
2	AD (PV)	0.855	0.8672	0.8869	0.9	0.8868
3	LPBE (PV)	0.905	0.9036	0.9314	0.9	0.9118
4	ABE (ART)	**0.9083**	0.9036	0.96	0.8666	0.9090

Table 6. 2-level based malignancy performance using support vector classifier

Iteration	Greedy feature	AUC	ACC	PRE	REC	F1
1	BC (PV)	0.8707	0.8472	0.8278	0.9333	0.8727
2	AD (ART)	0.8973	0.8290	0.8409	0.8761	0.8531
3	LPBE (PV)	**0.9040**	0.8290	0.8409	0.8761	0.8531

After using greedy strategy based feature selection, we can draw some conclusions from the results. For AUC performance target, gradient boosting classifier performs best. For a certain model, when the AUC score reaches the highest according to the greedy strategy, other scores may not be the best but is close to the highest one. From the results, we find that some features are good enough that they are selected by most of the popular models based on the greedy strategy. We summarize excellent features in Table 7.

Table 7. Detailed excellent features

Excellent feature	Perspective	Description
BC (PV)	Boundary	Chosen by 4 models based on greedy strategy
AD (PV)	Density	Chosen by 3 models based on greedy strategy
LPBE (PV)	Enhancement	Chosen by 3 models based on greedy strategy
WO	Enhancement	LI-RADS feature
C	Density	LI-RADS feature

Table 8. Comparison of the models based on the same feature set

Popular model	AUC	ACC	PRE	REC	F1
Support vector machine	0.8480	0.8290	0.8409	0.8761	0.8531
Decision tree	0.8383	0.8472	0.8778	0.8666	0.8613
Random forest	0.8454	0.7909	0.8314	0.8095	0.8172
Gradient boosting	**0.9483**	0.8654	0.8778	0.9	0.8831

For further comparison of the four popular machine learning models, we repeat the experiment based on the feature set listed in Table 7. The results are shown in Table 8.

4   Conclusion and Future Work

We proposed a set of computerized features for LI-RADS based computer-aided classification. As a preliminary study, we contacted 2-level (benign and malignant) classification experiments to validate the effectiveness of our proposed features using four popular machine learning models. The performance of each method was evaluated by using ROC curve and the best AUC score was 0.965 reached by the gradient boosting classifier. We will increase the data set for better analysis and performance of the model. There are still auxiliary features described in LI-RADS. They will be computerized in the future work. Also, level refactoring with more levels will be carried out in the future work.

Acknowledgement. This research was supported in part by the National Key Basic Research Program of China (973 Grant No. 2015CB352400), in part by the National Key Research and Development Program of China under the Grant No. 2016YFB1200203-03, in part by the Recruitment Program of Global Experts (HAIOU Program) from Zhejiang Province, China, in part by the Grant-in Aid for Scientific Research from the Japanese Ministry for Education, Science, Culture and Sports (MEXT) under the Grant No. 15H01130, No. 15K00253 and No. 16H01436.

References

1. An, C., Rakhmonova, G., Choi, J.Y., Kim, M.J.: Liver imaging reporting and data system (LI-RADS) version 2014: understanding and application of the diagnostic algorithm. J. Clin. Mol. Hepatol. 22(2), 296–307 (2016)
2. The American College of Radiology. https://nrdr.acr.org/lirads/
3. Mitchell, D.G., Bruix, J., Sherman, M., Sirlin, C.B.: LI-RADS (liver imaging reporting and data system): summary, discussion, and consensus of the LI-RADS management working group and future directions. J. Hepatol. 61(3), 1056–1065 (2014)
4. Ehman, E.C., Behr, S.C., Umetsu, S.E., Fidelman, N., Yeh, B.M., Ferrell, L.D., Hope, T.A.: Rate of observation and inter-observer agreement for LI-RADS major features at CT and MRI in 184 pathology proven hepatocellular carcinomas. J. Abdom. Radiol. 41(5), 963–969 (2016)
5. Clark, T.J., Mcneeley, M.F., Maki, J.H.: Design and implementation of handheld and desktop software for the structured reporting of hepatic masses using the LI-RADS schema. J. Acad. Radiol. 21(4), 491–506 (2014)
6. Shan, J., Alam, S.K., Garra, B., Zhang, Y., Ahmed, T.: Computer-aided diagnosis for breast ultrasound using computerized BI-RADS features and machine learning methods. J. Ultrasound Med. Biol. 42(4), 980 (2016)
7. Jha, R.C., Mitchell, D.G., Weinreb, J.C., Santillan, C.S., Yeh, B.M., Francois, R., Sirlin, C.B.: LI-RADS categorization of benign and likely benign findings in patients at risk of hepatocellular carcinoma: a pictorial atlas. Am. J. Roentgenol. 203(1), 48–69 (2014)
8. Khan, A.S., Hussain, H.K., Johnson, T.D., Weadock, W.J., Pelletier, S.J., Marrero, J.A.: Value of delayed hypointensity and delayed enhancing rim in magnetic resonance imaging diagnosis of small hepatocellular carcinoma in the cirrhotic liver. J. Magn. Reson. Imaging 32(2), 360–366 (2010)
9. Marrero, J.A., Hussain, H.K., Nghiem, H.V., Umar, R., Fontana, R.J., Lok, A.S.: Improving the prediction of hepatocellular carcinoma in cirrhotic patients with an arterially-enhancing liver mass. J. Liver Transplant. 11(3), 281–289 (2005)
10. Bishop, C.M.: Pattern Recognition and Machine Learning. Springer, New York (2006)

Improving Measurement Accuracy for Rheumatoid Arthritis Medical Examinations

Tomio Goto[1,2]($\boxtimes$), Yoshiki Sano[1,2], Takuma Mori[1,2], Masato Shimizu[1,2], and Koji Funahashi[1,2]

[1] Department of Computer Science and Engineering, Nagoya Institute of Technology, Gokiso-cho, Showa-ku, Nagoya 466-8555, Japan
t.goto@nitech.ac.jp, {shiki,mori,asimo}@splab.nitech.ac.jp
[2] Plastic Surgery, Kariya Toyota General Hospital,
5-15 Sumiyoshi-cho, Kariya, Aichi 448-8505, Japan
http://www.splab.nitech.ac.jp/

Abstract. Super-resolution techniques have been widely used in fields such as television, aerospace imaging, and medical imaging. In medical imaging, X-rays commonly have low resolution and a significant amount of noise, because radiation levels are minimized to maintain patient safety. So, we proposed a novel super-resolution method for X-ray images, and a novel measurement algorithm for treatment of rheumatoid arthritis (RA) using X-ray images generated by our proposed super-resolution method. In this paper, we improve measurement accuracy for our proposed method. Moreover, to validate it for our proposed algorithm, we make a model for measurement algorithm about joint space distance using a 3D printer, and X-ray images are obtained to photograph it. Experimental results show that high quality super-resolution images are obtained, and the measurement distances are measured with high accuracy. Therefore, our proposed measurement algorithm is effective for RA medical examinations.

Keywords: Super-resolution · Joint space distance · Rheumatoid Arthritis · Medical examinations

1 Introduction

X-ray images are widely used to diagnose a variety of diseases. However, to reduce the patient's exposure to radiation, X-ray dosage is minimized as much as possible. As a result, X-ray images contain a significant amount of noise and resolution is compromised. Thus, it is necessary to increase image resolution and reduce noise. We proposed a novel super-resolution system for X-ray images that consists of total variation (TV) regularization, a shock filter, and a median filter. In addition, we proposed a novel measurement algorithm for the treatment of RA, using X-ray images generated by our proposed super-resolution system [1]. In this paper, we improve measurement accuracy to optimize parameters for super-resolution.

© Springer International Publishing AG 2018
Y.-W. Chen et al. (eds.), *Innovation in Medicine and Healthcare 2017*, Smart Innovation,
Systems and Technologies 71, DOI 10.1007/978-3-319-59397-5_17

2   Super-Resolution System

Super-resolution is a technique for increasing the resolution of an enlarged image by generating new high-frequency components. This technique estimates and generates such components from the characteristics of the original signals. In recent years, various super-resolution techniques have been proposed, and most are classified as either reconstruction-based super-resolution [2] or learning-based super-resolution [3].

In resolution for X-ray images, which is used at medical operation, the resolution of one pixel will be 0.15 mm square, and joint space distance will be about from 1.0 mm to 1.5 mm. Therefore, when the joint space distance is measured with one pixel error, the error will be from 10% to 15%, improving measurement accuracy is required.

It is possible to solve this problem by utilizing a super-resolution technique. By magnifying segmented images by a multiple of 4 × 4, a resolution of a pixel will be 0.0375 mm, thus the error can be suppressed from 2.5% to 3.75%. Also by utilizing the shock filter, clearer edges in images will be obtained, thus it is easy to select edges. However, the error still remains and it is not small to measure for RA medical examinations. In this paper, we improve measurement accuracy to set bigger magnification rate and optimal parameters for super-resolution.

A block diagram of our proposed super-resolution system is shown in Fig. 1. Each of the non-linear filters is explained in the following sections.

2.1   Total Variation Regularization

As shown in Fig. 1, the TV regularization decomposition [4–8] is performed as follows. The structure component u is calculated to minimize the evaluation function $F(u)$ as shown in Eq. (1):

$$F(u) = \Sigma_{i,j}|\nabla u_{i,j}| + \lambda\Sigma_{i,j}|f_{i,j} - u_{i,j}|^2. \tag{1}$$

where f is a pixel value of the input image. The Chambolle's projection algorithm [9] is used to solve the minimization problem as shown in Eq. (2).

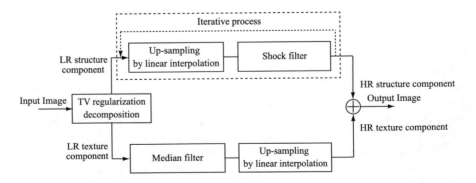

Fig. 1. Block diagram of our proposed super-resolution system.

$$P_{i,j}^{(n+1)} = \frac{P_{i,j}^{(n)} + (\frac{\tau}{\lambda})\nabla(f + \lambda div P_{i,j}^{(n)})}{max\{1, |P_{i,j}^{(n)} + (\frac{\tau}{\lambda})\nabla(f + \lambda div P_{i,j}^{(n)})|\}}. \tag{2}$$

where P is a pixel value. The texture component v and the structure component u are obtained by using the equations in (3).

$$v = \lambda div P, \qquad u = f - v. \tag{3}$$

Figure 2 shows an example of the TV decomposition for an X-ray image, (a) is an original image, (b) is a structure component and (c) is a texture component.

2.2 Shock Filter

The shock filter is a nonlinear edge enhancement filter, which was proposed by Osher and Rudin [10] and Alvarez and Mazorra [11]. The process is achieved by utilizing Eq. (4):

$$u_{i,j}^{(n+1)} = u_{i,j}^{(n)} - sign\left(\Delta\left(K_\sigma * \Delta u_{i,j}^{(n)}\right)\right)\left|\nabla u_{i,j}^{(n)}\right| dt. \tag{4}$$

where u is a structure component, K is a smoothing filter. It is possible to reconstruct steep edges by calculating a simple operation; thus, this filter is suitable for high-speed processing. In addition, several artifacts generated during edge enhancement processing can be controlled successfully, i.e., ringing noise and jaggy noise. Figure 3 shows an input image and output images obtained by utilizing the shock filter.

2.3 Median Filter

Most noise is classified as a texture component by utilizing TV regularization. As mentioned previously, X-ray images contain significant noise. Therefore, we propose applying the median filter to the texture components of X-ray images. The median filter sorts nine pixel values in 3 × 3 pixels around the pixel of interest. Next, the filter replaces the fifth pixel value with a new pixel value of interest. This process is applied to all of the texture components.

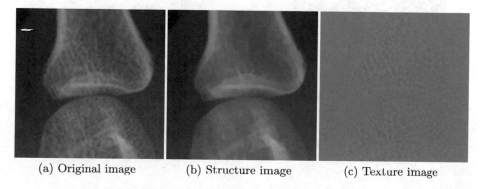

(a) Original image   (b) Structure image   (c) Texture image

Fig. 2. Example of total variation decomposition for X-ray image.

3   Measurement Algorithm About Joint Space Distance

Rheumatoid arthritis (RA) is a disease that causes joint inflammation, and most commonly afflicts women between 30 and 50 years of age. As symptoms progress, the patient's joint space distance (JSD) will narrow. This change can be observed with X-ray images; however, at present, an accurate measurement method has not been established. Therefore, a more accurate JSD measurement technique is required. We proposed two JSD measurement algorithms [1]. Figure 4 shows an example of an output image. In our proposed method, we use an input image, which is magnified by utilizing a super-resolution method. Our algorithm for measurement is shown as follows:

1. Select several points on an edge of an upper bone by clicking mouse button, manually.
2. Set axes corresponding to a joint from selected points automatically.
3. Calcurate coordinate values of selected points based on the axes.
4. Calcurate a quadratic function by using the least squares method from the calcurated fitting function.
5. Calcurate a quadratic function by selecting several points on an edge at a lower bone, similarly.
6. Measure the joint space distance from normal lines and integral operation.

We also propose another JSD measurement algorithm, which is calcurated by using an area of JSD, and its algorithm for measurement is shown as follows:

1. An integral calculus range in a range of curve p is set, and to calculate the integral calculus S_R of differences between curve p and curve q.
2. An integral calculus range in a range of curve q is set, and to calculate the integral calculus S_B of differences between curve p and curve q.
3. The value W is defined as the distance between 2 points, which is automatically detected when the coordinate axis are set.

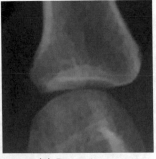
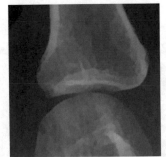

(a) Input image   (b) Output image from shock filter

Fig. 3. Example of shock filter for X-ray image.

4. The JSD value D is calculated from two values: S_R and S_B and averages of the distances of the top and bottom as shown in Eq. (5).

$$D = \frac{S_R + S_B}{2W} \tag{5}$$

4 Experimental Results

To validate measurement accuracy in our proposed method, we experiment using 3D modeled images.

4.1 3D Modeled Images

We have designed the 3D model in imitation of the joint of the left hand middle finger, which has three joint space distances: 1 mm, 2 mm and 3 mm as shown in Fig. 5, and have printed out it by using a 3D printer. Then, an X-ray image is obtained to photograph it, we compare the distances between the ideal one and measured one by utilizing our proposed application. The measurement distances are calcurated at a mean value and an area by a distance by using Eq. (5).

The experimental condition is shown in Table 1, the model made by a 3D printer and its X-ray image are shown in Figs. 6 and 7, respectively.

4.2 Super-Resolution System

Figure 8 shows (a) original image, the experimental results magnified (b) twice, (c) 4 times and (d) 8 times for the super-resolution system. In Fig. 8, the original

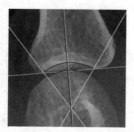

Fig. 4. Application for JSD calculation of X-ray image.

Table 1. Experimental parameters.

Parameter	Value
Width of joint space distance	3 [mm]
Distance from X-ray to objects	1000 [mm]
Magnified rate	1.5

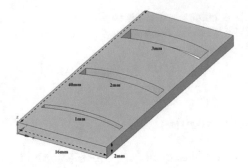

Fig. 5. Designed 3D model

image is very small area, and the magnification rate is bigger, the clearer edges are obtained.

We will measure the JSD 50 times by using four X-ray images — an input image before the super-resolution, the linear interpolation images and the super-resolution images magnified twice, 4 times and 8 times. Table 2 and Fig. 9 show measurement results for X-ray images of printed 3D model. In Table 2 and Fig. 9, when the bigger magnification rate is set, the smaller error rate is obtained, and the distribution is getting smaller. Also, the measurement accuracy of the average JSD for our super-resolution system is higher than that for the linear interpolation method in each magnification rate. As a result of 8 times magnification rate, the error rate is 0.77%, this means that the measurement accuracy is very high quality for our proposed super-resolution system, and it is possible to verify and monitor the earlier state of RA, objectively.

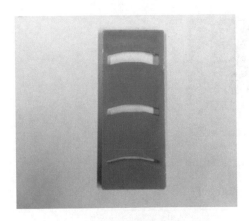

Fig. 6. Printed 3D model

Fig. 7. X-ray image of printed 3D model.

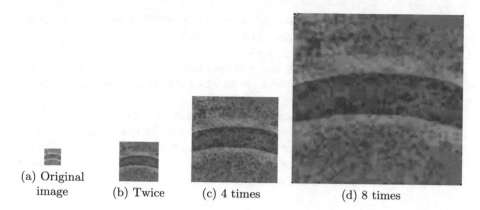

(a) Original
image   (b) Twice   (c) 4 times   (d) 8 times

Fig. 8. Original image and experimental results of super-resolution.

Table 2. Measurement results for X-ray images of printed 3D model.

Magnification rate	Original	Linear interpolation			Super-resolution		
		2 times	4 times	8 times	2 times	4 times	8 times
Max. JSD [mm]	3.26	3.259	3.215	3.188	3.132	3.099	3.087
Min. JSD [mm]	2.88	2.965	2.958	2.966	2.941	2.948	2.961
Max.-Min. JSD [mm]	0.38	0.294	0.257	0.222	0.191	0.151	0.126
Average JSD [mm]	3.15	3.120	3.092	3.081	3.084	3.035	3.023
Error rate [%]	5.00	4.00	3.07	2.70	2.80	1.17	0.77

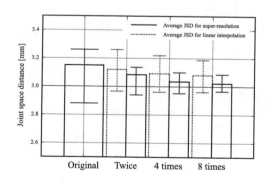

Fig. 9. Measurement results

5   Conclusion

In this paper, we have validated measurement accuracy for our proposed super-resolution system for X-ray images that utilizes the TV regularization, the shock filter, and the median filter. Experimental results show that high-quality super-resolution images are obtained from low-quality X-ray images utilizing our proposed super-resolution system. In addition, the results show that JSD can be measured more accurately and the progress of RA can be monitored more precisely by using super-resolution images.

Also, we made a model by using a 3D printer to validate our measurement algorithm, and by measuring it, we have validate that measurement accuracy of our proposed method is very high quality.

For further research, we intend to measure JSDs automatically because several points' selection on edges of upper and lower bones by users is required in our proposed measurement algorithm, and it may include error. Also, we intend to improve the user interface of our proposed measurement system, which doctors use in medical examination and treatment.

References

1. Goto, T., Mori, T., Kariya, H., Shimizu, M., Sakurai, M., Funahashi, K.: Super-resolution technology for X-ray images and its application for rheumatoid arthritis medical examinations. In: International KES Conference on Innovation in Medicine and Healthcare (InMed), pp. 217–226, June 2016
2. Singh, A., Ahuja, N.: Single image super-resolution using adaptive domain transformation. In: IEEE International Conference on Image Processing (ICIP), pp. 947–951, September 2013
3. Cho, C., Jeon, J., Paik, J.: Example-based Super-resolution using selfpatches and approximated constrained least squares filter. In: IEEE International Conference on Image Processing (ICIP), pp. 2140–2144, October 2014
4. Rudin, L., Ohser, S., Fetami, E.: Nonlinear total variation based noise removal algorithm. Phys. D **60**, 259–268 (1992)
5. Meyer, Y.: Oscillating Patterns in Image Processing and Non-linear Evolution Equation, the fifteenth Dean Jacqueline B. Lewis Memorial Lectures, American Mathematical Society, University Lecture Series, Vol. 22 (1992)
6. Vese, L.A., Osher, S.J.: Modeling texrures with total variation minimization and oscillating patterns in image processing. J. Sci. Comput. **19**, 553 (2003)
7. Aujol, J.-F., Aubert, G., Blanc-Feraud, L., Chambolle, A.: Image decomposition into a bounded variation component and an oscillating component. J. Math. Imaging Vis. **22**, 71–88 (2005)
8. Goto, K., Nagashima, F., Goto, T., Hirano, S., Sakurai, M.: Super-resolution for high resolution displays. In: IEEE Global Conference on Consumer Electronics (GCCE), pp. 309–310, October 2014
9. Chambolle, A.: An algorithm for total variation minimization and applications. J. Math. Imaging Vis. **20**(1), 89–97 (2004)
10. Osher, S.J., Rudin, L.I.: Feature-oriented image enhancement using shock filters. SIAM J. Numer. Anal. **27**, 910–940 (1990)
11. Alvarez, L., Mazorra, L.: Signal and image restoration using shock filters and anisotropic diffusion. SIAM J. Numer. Anal. **31**(2), 590–605 (1994)

Design and Implementation of a Social Networking Service-Based Application for Supporting Disaster Medical Assistance Teams

Toshiki Kawai[1]([✉]), Haruka Kambara[1], Kohei Matsumura[1], Haruo Noma[1],
Osamu Sugiyama[2], Manabu Shimoto[2], Shigeru Ohtsuru[2],
and Tomohiro Kuroda[2]

[1] College of Information Science and Engineering,
Ritsumeikan University, Kyoto, Shiga, Japan
tkawai@mxdlab.net
[2] Kyoto University Hospital, Kyoto, Japan

Abstract. During the Kumamoto earthquakes in Japan, disaster medical assistance teams (DMATs) were dispatched for emergency support. Communication among DMAT members were primarily done via emails and phones, however, during this disaster, some teams also used LINE, a popular social networking service in Japan. Although this tool is simple to use, the teams had problems organizing various topics in a single chat room. In this paper, we propose an application that uses hashtags, which consists of two main units: (1) a bot that redirects messages to specific groups according to hashtags input by users; and (2) a system for logistic-support to manually apply hashtags to messages without tags, and to manually edit hashtags of already-tagged messages. User studies of two Kyoto University Hospital DMAT members were conducted, and through discussion, we found that the generality of the proposed application should be further considered for usage in other activities.

Keywords: DMAT · EMIS · SNS · LINE · Tags

1 Introduction

On April 2016, Kumamoto Prefecture, Japan was hit by a series of earthquakes, causing many casualties. Disaster medical assistance teams (DMATs), medical teams dispatched in cases of emergencies, were sent to give medical support from all over Japan. A DMAT is split into two main groups, a group on-site, and a logistic-support group that stays at their base hospital.

DMATs primarily used the Emergency Medical Information System (EMIS) as a communication tool. The EMIS is a system that holds and shares information between institutions, including other DMATs, hospitals and administrations. DMAT members communicate with other institutions by inputting

Y.-W. Chen et al. (eds.), *Innovation in Medicine and Healthcare 2017*, Smart Innovation, Systems and Technologies 71, DOI 10.1007/978-3-319-59397-5_18

information into the EMIS, such as their current activity, vacancy of hospitals, and conditions of infrastructure.

In contrast, conventional communication methods among DMAT members have been via phone and email. However, for the Kumamoto earthquakes, some DMATs also used LINE, a social networking service (SNS) that uses chat rooms, as a communication tool for the first time. LINE is a popular SNS in Japan used by over 40% of the population in 2014 [5]. Its simplicity and familiarity provide DMATs members smooth communication. To differentiate the purposes between EMIS and LINE, Fig. 1 shows the flow of disaster information during the Kumamoto earthquakes. The EMIS was used to share information between different organizations, and LINE was used for communication among DMAT members.

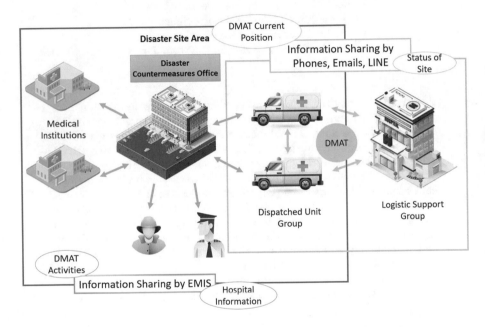

Fig. 1. Flow of information during the Kumamoto earthquakes

Although LINE is effective for daily use, this has caused other problems when using it during emergency situations. A primary problem was: excessive amount of messages in a single chat room causing crucial information to be left unseen. Missing important messages is tragic, therefore, in this paper, we propose an application to organize various messages.

2   Method

2.1   Problem Analysis

For further understanding and clarification of the problem caused by large amounts of messages, de-identified log files of the Kyoto University DMAT chat room were analyzed. There were no problems regarding how to use LINE because members had already been using LINE, however, there were too many topics, including instructions and small talk, going on at the same time in a single chat room. This forced members to search for information that was related or important to them. Having multiple topics led to more difficulties, e.g., members were required to distinguish between to which topic a certain message was referring. To make matters worse, according to the log files, only a mere five percent of all messages were the most crucial messages, instructions, which must not be missed.

2.2   Solution Considered

As a solution for this problem, we considered implementing a system for distributing messages to different chat rooms using a chat bot, as shown in Fig. 2. We employed LINE for this system because LINE is more widely used than other communication tools, such as Slack. Slack is a tool that may be functionally better, but is mainly used by IT specialists. DMAT members are already experienced in the basic uses of LINE, and will only need to learn how to use the additional features created.

Users will manually apply metadata, or data that describe other data, to their messages beforehand. Based on these metadata (henceforth referred to as "tags") placed in the messages, the bot will then classify them accordingly. Since instructions are crucial messages that DMAT members need to see, it is necessary for the sender of the instructions to confirm that the receivers viewed them. Therefore, receivers of the instructions should be able to reply to the instructions as a confirmation.

A bot is able to automatically categorize messages through language processing; however, members tend to send short messages along with occasional errors, which makes them difficult for the bot to analyze. This raises the possibility of incorrect categorizations, which is a major issue for DMATs where accurate information is in need. In addition, as members at the disaster site are preoccupied with their operations, misplacing and forgetting tags are unavoidable. Thus, the logistic support groups with more free time will apply forgotten tags or correct mistaken tags.

3   Proposed Application

The implemented application is made up of two main units: (1) a bot that distributes messages to different chat rooms according to their tags; and (2) a support system for logistics to fix tag-related errors. The former unit prevents the problem of significant messages being missed inside a single integrated chat room. The latter unit corrects human errors when attaching tags to messages. There is a one-to-one chat room with the bot for each user, a group chat room for each tag, and a web page for the logistic support system. The former unit utilizes the hashtag function from Twitter due to structural limitations of LINE. For this application, "EMIS", "Instruction", "Transportation", "Activity", and "Others", were set as suitable tags.

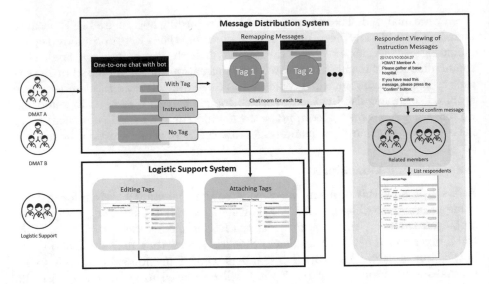

Fig. 2. Concept of application

3.1   Message Distribution Bot

This bot consists of two features: a) remapping messages to different chat rooms according to the attached tags; and b) viewing the respondents of an instruction-related message sent by the user.

Users select tags, as shown in Fig. 3, before sending the actual message. After sending the message, the bot will distribute the message to the chat rooms of the specified tags. In addition, users can also check messages of all chat rooms.

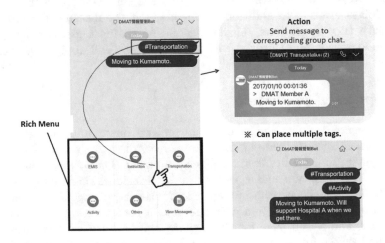

Fig. 3. Selecting tags and sending messages

When a user sends a message with an "Instruction" tag, the bot sends a confirm message to each user, as shown in Fig. 4. If a receiver presses "Yes", he is considered a viewer, or a respondent to the message. Then, as shown in Fig. 5, the sender can view all respondents. The sender will now be able to confirm that the instructions sent have been viewed.

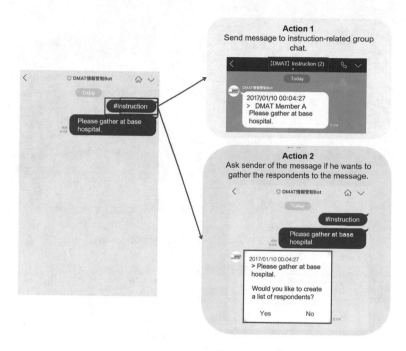

Fig. 4. Confirm messages for instructions

The remapping feature enables users to view messages by categories instead of viewing all messages at once, hence, preventing confusion from reading multiple topics in a single chat room simultaneously. The reallocate confirmation feature provides confirmation to the sender that the crucial instructions sent have been viewed, and visual understanding of the receivers of the message being important.

3.2 Support System for Logistics

This system consists of two features: (a) attaching tags to messages without any tags; and (b) revising the tags of already-tagged messages. The web page consisting of this system is divided into two sections, as shown in Fig. 6. The left shows the list of messages without tags, and the right shows messages of all chat rooms.

Users apply tags to a message with none, as shown in Fig. 7, by simply clicking on one of the messages, selecting the tags, and confirming it. The message will then be sent to the respective chat rooms.

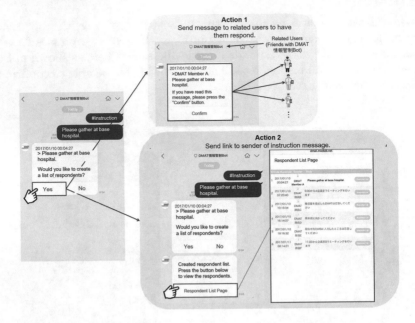

Fig. 5. Viewing instruction respondents

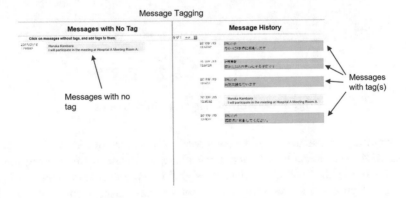

Fig. 6. Logistic support system

Likewise, users select an already-tagged message, and add or remove tags, as shown in Fig. 8. The message will be resent to the respective chat rooms.

This system corrects any messages without tags and messages with misplaced tags, to properly send messages to their appropriate chat rooms, and to prevent misleading understandings by members respectively. The logistic-support group with more free time will manually revise the tags to fix the human errors caused by the preoccupied members at the disaster site.

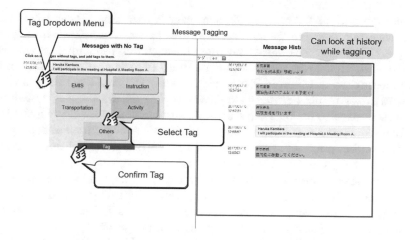

Fig. 7. Attaching tags to messages without tags

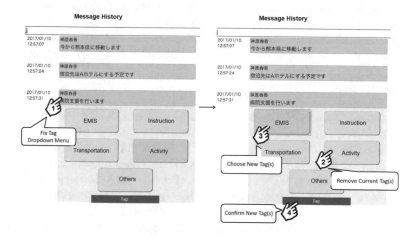

Fig. 8. Revising tags

4   Discussion and Improvements

User studies of two Kyoto University Hospital DMAT members, who were active during the Kumamoto earthquakes, were conducted after designing the application. One member was part of the logistic-support team, and the other was the leader of the dispatched team. Both members are experienced in DMAT activities and instructs other members during training. The members were asked to use the prototype application for a week, and be interviewed about each application feature through a free discussion. During the discussions with the users, we obtained feedback and suggestions for additional features for the application.

There were various improvement points suggested for the message-distribution system. One user stated, *"The tags are too focused on disasters; therefore, the system should have more generality to be usable for other emergency situations.*" Another comment from a user was, "*Placing tags should be possible after inputting the message, because it is easier to decide the tags after writing it.*" Because there was a chat room for each tag, users found it confusing, and suggested that fewer chats would be more beneficial. For the respondent-viewing feature, a user stated, "*It would be better if I could choose who to send the instructions to, and if there was more variety in the ways of responding.*" Users also wanted to manage the status of respondents, to see if the instructions were followed. Being able to determine a member's availability was suggested as well. If a member does not send a message for a certain period of time, a message should be sent to confirm his current status.

For the logistic-support system, the web page was not suitable for smartphones, which was one of the comments given. Multiple comments about the user interface were given, such as, "*Selection of colors is bad.*", "*It should look more like LINE.*". These should be addressed immediately.

To summarize, LINE was first used by some DMATs because of its simplicity and familiarity. Given that most users are already experienced in using LINE, the proposed application is easier to use than other communication tools such as Slack. However, the application caused difficulties during use. For it to be usable in various emergency situations, considerations of the design to be generalized and to have the same simple user interface that LINE has must be made.

5   Conclusion

We proposed an application using LINE that consists of two units to support communications among DMAT members. One unit consists of a bot that automatically distributes messages to chat rooms according to the tags the user has attached to the messages; and the other unit for logistic-support members, allows adding tags to messages without tags and revising tagged messages. Through discussion and feedback, various improvement points were considered for both units, and are currently under development. For future work, considerations for implementing this application in actual DMAT training are being made for further discussion. In addition, there was a problem of members forgetting to input information to the EMIS upon sending the same information through LINE. Therefore, a new system to support the coordination of information between the EMIS and LINE is currently being considered.

References

1. What is DMAT? DMAT Homepage. http://www.dmat.jp/DMAT.html
2. Japanese Association for Disaster Medicine. DMAT Text Revision Editorial Committee (2015). "2nd Edition DMAT Standard Textbook". Published by Herusu Shuppan Co. Inc.
3. Heisei 28 1st Kyoto DMAT Liaison Council (Heisei 28 Kyoto DMAT Activity Report of Kumamoto Earthquakes)
4. LINE API Reference. https://devdocs.line.me/en/
5. LINE is used by 40% of the population, but Facebook Messenger and WhatsApp are the mainstream apps in the world. Ascii.jp. http://ascii.jp/elem/000/001/016/1016194/

Quantitative Assessment of Small Bowel Motility Using Cine MR Sequence Images and Superpixels

Tomoko Tateyama[1,2(✉)], Ayako Taniguchi[2], Akira Furukawa[3],
Makoto Wakamiya[4], Shuzo Kanasaki[5], Kazuki Otsuki[2],
and Yen-Wei Chen[2]

[1] Faculty of Applied Information and Science,
Hiroshima Institute of Technology, Hiroshima, Japan
t.tateyama.es@cc.it-hiroshima.ac.jp
[2] College of Information Science and Engineering,
Ritsumeikan University, Kusatsu, Japan
[3] The Graduate School of Human Health Sciences,
Tokyo Metropolitan University, Tokyo, Japan
[4] Ragiology, Koseikai Takeda Hospital, Kyoto, Japan
[5] Ragiology, Nagahama City Hospital, Shiga, Japan

Abstract. In this paper, we propose a method for automated assessment of small bowel motility function based on image analysis using Cine MR Image. In this study, we first use a superpixels based method to segment the target small bowel region and then the temporal area change of the segmented small bowel region is used for quantitative assessment of the small bowel motility function. The main contribution of this paper is to improve the measurement accuracy of the small bowel motility function from Cine MR imaging and develop an efficient, useful and automatic assessment system based on image processing.

Keywords: Cine MR images · Assessment of small bowel motility function · SLIC superpixel · Measurement dynamic change of region area

1 Introduction

Measurement of small bowel contraction movement is important for the treatment or inspection of the small intestine. To date, the popular measurement is an invasive method using some endoscope. In our previous work [1], we proposed a non-invasive method using Cine MR images. Since Cine MR images can provide temporal sequential images, we are able to observe small intestine contraction movement directly.

In order to make an automatic assessment of small bowl motility, we proposed several methods based on image processing and computer vision techniques [2–4]. The basic idea is that we first select a small intestine as a target object and then automatically segment it using computer vision techniques. The temporal areas change of the segmented small intestine is used as a measure for quantitative assessment of small bowel motility. Many methods have been proposed for medical image segmentation,

© Springer International Publishing AG 2018
Y.-W. Chen et al. (eds.), *Innovation in Medicine and Healthcare 2017*, Smart Innovation,
Systems and Technologies 71, DOI 10.1007/978-3-319-59397-5_19

such as Otsu (discriminant analysis) method [5], watershed [6] and level-set [7], etc. Otsu-method is used to automatically find an optimum threshold to separate the object from the background. However, it cannot be used for images having complex background. Watershed and level-set methods need good initial point or boundary. On the other hand, superpixels based segmentations received great attentions. Simple Linear Iterative Clustering, called SLIC [8], is one of superpixels methods, which clusters pixel using the combined color (or feature) distance and spatial distance. In this paper, we propose a superpixels based segmentation method to segment the small intestine and do quantitative assessment of small bowel motility compared with existing methods Otsu method and active contour methods, our proposed method can achieve better results.

This paper is organized as follows. Section 2 describes the detail of our clinical material, such as Cine MR imaging and clinical preparation. Our proposed assessment system with superpixels is presented in Sect. 3. The experimental results of proposed method and conclusion are demonstrated in Sect. 4.

2   Cine MR Images

Dynamic Cine MR imaging is a useful tool to observe the bowel contraction and provide functional information of the bowel. The Cine MR images we used in this research were provided by Shiga University of Medical School and were collected by a 1.5T MR scanner (Signa HDxt 1.5T, GE Healthcare) with an 8-channel body array coil. We use coronal images of the entire abdomen to observe the small bowel contraction because it covers the maximum length of smack bowel loops. The coronal plane was selected by consensus of two radiologists. A serial coronal scan was obtained at the selected plane with the tested subject in a prone position: 70 images in 30 s with breath-holding or 200 images in 90 s without breath-holding. Steady-state free precession sequence (FIESTA Sequence: TR 1/4 3.4 ms, TE 1/4 1.2 ms, flip angle 75 degree, slice thickness 10 mm, matrix 256×256, field of view 450 mm) was utilized for imaging, which allows continuous scanning at every 0.5 s.

The Cine MR images used in this research were taken from 4 males healthy volunteers (33–49 years) as summarized in Table 1. Before 20 min prior to initiation of scanning, the volunteers drank 1500 (ml) of non-absorbable fluid to obtain optimal distention of the small bowel loops for easier recognition of their walls and accurate measurement of their caliber on each MR images. The sequential scanning was repeated before and at 0, 15, 30, 45 and 60 min after oral intake of the non-absorbable fluid. In Fig. 1, we show examples of Cine MR sequential images demonstrating bowel contraction with an interval time of 0.5 s.

Table 1. Experimental clinical datasets

Sample No.	Sex	Area	Elapse time (min)
103	Male	LU, RL	0, 15, 30, 45, 60
104	Male	LU, RL	0, 15, 30, 45, 60
105	Male	LU, RL	15, 30, 45
106	Male	LU, RL	15, 30, 45

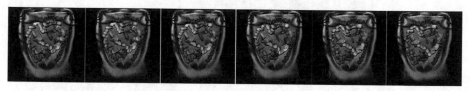

Fig. 1. Cine MR sequential images showing bowel contraction with an interval time of 0.5 s

3 Assessment of Small Bowel Motility with Cine MR Sequence

3.1 System Overview

The proposed assessment system of small bowel motility function with Cine MR sequence is shown in Fig. 2. The assessments were performed quantitatively by measuring frequencies of bowel contraction or by measuring the signal changes between sequential images [1]. The system consists of 3 steps. The first step is to define a region of interest (ROI) for analysis. The radiologist marks two points: left-up and right-down points of a rectangle area (ROI) manually as shown in Fig. 3. The second step is the segmentation or measurement step. In this step, we automatically segment the target small intestine in the ROI of each frame by using image processing techniques, which will be described in next sub-sections, in order to measure the size or area of the small intestine. In the final step, we quantitatively analysis the small bowel motility function by measuring the temporal area change of the segmented small intestine. The frequency analysis is also done based on Fourier transformation.

3.2 Superpixel-Based Segmentation of Small Intestine

In our previous work, the radiologist measured the luminal diameter of each target small intestine manually [1]. It is a time-consuming and labor-intensive task. To overcome the problem, we propose an automatic segmentation of small intestine based on a superpixel technique.

Superpixel can be considered as a clustering method, which clusters similar neighboring pixels into superpixels. Simple Linear Iterative Clustering, called SLIC, is one of superpixel algorithms, which is based on K-means algorithm. K-means clusters

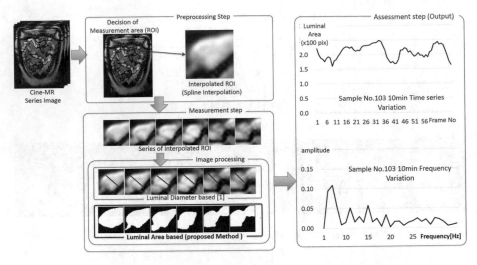

Fig. 2. Overview of the assessment system of small bowel motility with Cine MR images

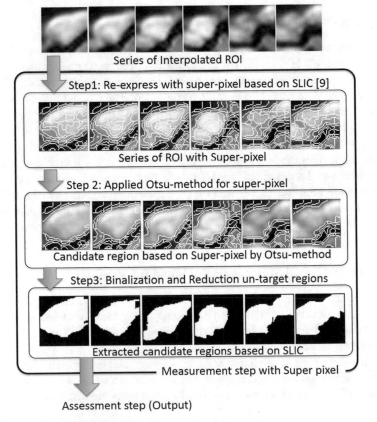

Fig. 3. Flow of the superpixel-based small intestine segmentation experimental results

pixels using only color (or feature) distance, while SLIC clusters pixels using the combined color (or feature) distance and spatial distance. Since the MR image does not have color information, we use intensity I as a feature instead of color values. The similarity measure is based on following two distance measures:

$$d_c = \sqrt{(I_c - I_i)^2},$$ (1)

$$d_s = \sqrt{(x_c - x_i)^2 + (y_c - y_i)^2},$$ (2)

where d_c is the distances (difference) between the i-th pixel's intensity (I_i) and the c-th cluster center's intensity (I_c). d_s is the Euclidean distance between the i-th pixel (x_i, y_i) and the c-th cluster center (x_c, y_c). The combined distance measure D are shown in Eq. (3),

$$D = \sqrt{d_c^2 + (d_s/S)^2 m^2}.$$ (3)

where S is the grid interval and m is a control parameter, which controls the compactness of the superpixel. The greater the value of m, the more important the spatial distance of d_s resulting in more compact the cluster. If $m=0$, the SLIC can be considered as the conventional K-means clustering. In this study, we performed the SLIC by Ref. [8].

The flow of our superpixel-based small intestine segmentation is shown in Fig. 3. It consists of 3 steps. In the first step, we apply the SLIC algorithm to the ROI of each frame and represent them with superpixels. Since the small intestine (object) has higher intensity than background, In the second step, if the average intensity of a superpixel is larger than a pre-defined threshold, we relabeled it as an object and merged the neighboring object superpixels as one object superpixel since the small intestine (object) has higher intensity than background. The threshold was estimated by the use of Otsu method [5]. In the last step, only the object superpixel with the largest area is selected as the region of small intestine. The small object regions are deleted as noise.

To validate the effectiveness of the proposed method, several experiments were performed. We first manually segmented target regions under the guidance of medical doctors, which are used as ground truth. Figure 4 shows detailed comparisons of segmentation results. Figure 4(a) is an example of dynamic MR images (Patient:S103, Timestamp:45, Frame No. 1, 2, 3, 30, 70). Figures 4(b)–(e) are results of manual segmentation, proposed method (SLIC), level-set method, Otsu method, respectively. It can be seen that we cannot obtained a good segmentation result by the conventional level-set method and Otsu method, while our proposed superpixels based method performs well. The temporal area changes of the segmented small bowel region are shown in Fig. 5(a) and their Fourier transforms are shown in Fig. 5(b), which can be used for quantitative assessment of the small bowel motility function. The red line is measured by our proposed SLIC method. The black dotted line is the result of manual

178   T. Tateyama et al.

segmentation, which is used as the ground truth. The blue line is the result by conventional Otsu method and the orange dashed line is the result by level-set method. It can be seen that both the temporal area change and the Fourier transform result by our proposed method is most similar to the ground truth among three methods.

In order to make a quantitative comparison, we calculated the correlations between the manual segmentation result and the automatic segmentation results, which are summarized in Table 2. Our proposed method outperformed both Otsu method and level-set method. The computation time is shown in Fig. 6. Our proposed superpixels based method takes about 500 s. In the future, we are going to develop a fast superpixel algorithm.

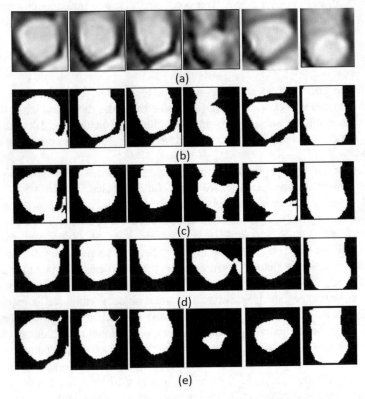

Fig. 4. Segmentation results of ROI series on Cine MR images, Sample No. 103 (Timestamp is 45 min) Frame No. 1, 2, 3, 30, 50, 70. (a) original ROI series from the clinical dataset, (b) manual segmentation, (c) segmentation by proposed method (SLIC superpixel), (d) segmentation by Otsu method and (e) segmentation by Level-set method.

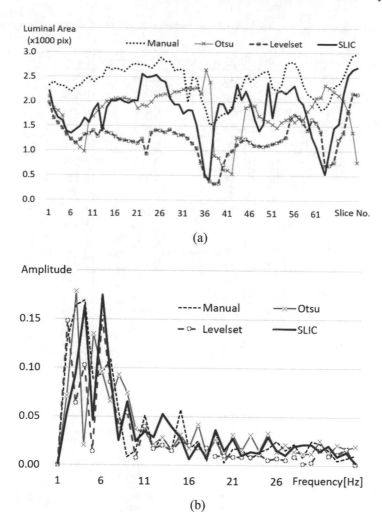

Fig. 5. Comparison results of temporal area changes of the segmented small bowel region (a) and their Fourier transforms (b). SLIC (Red), manually (black dot), Otsu Method (blue with symbol x), level-set (orange line with circle). Sample is No. 103 with Timestamp 45 min)

Table 2. Correlations between the manual segmentation result and the automatic segmentation results

Sample No.	Min.	SLIC (proposed)	Otsu-method	Level-set
S103	10	0.848	0.809	0.763
	15	0.818	0.762	0.424
	30	0.598	0.486	0.377
	45	0.745	0.563	0.626
	60	0.600	0.636	0.735
S104	10	0.517	0.392	0.269
	15	0.596	0.799	0.424
	30	0.654	0.562	0.050
	45	0.725	0.005	0.311
	60	0.355	0.250	0.881
S105	15	0.911	0.962	0.767
	30	0.771	0.931	0.310
	45	0.190	0.529	0.466
S106	15	0.724	0.542	0.381
	30	0.504	0.404	0.095
	45	0.628	0.732	0.545
Total average		0.636	0.585	0.464

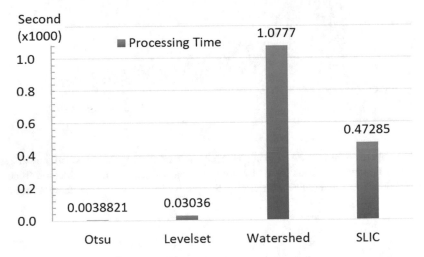

Fig. 6. Comparison of computation time.

4   Conclusion

In this paper, we proposed a method for automated assessment of small bowel motility function based on image analysis using Cine MR Image. We first used a superpixel-based method to segment the target small bowel region and then the temporal area change of the segmented small bowel region is used for quantitative assessment of the small bowel motility function. Compared with existing methods such as Otsu method and level-set method, our proposed method can achieve better results.

Acknowledgement. This work is supported in part by the Grant in Aid for Scientific Research from the Japanese Ministry for Education, Science, Culture and Sports (MEXT) under the Grant Nos. 15K16031, 25461839, 15H01130 and 24103710; in part by the MEXT Support Program for the Strategic Research Foundation at Private Universities, Grand No. S1311039 (2013–2017), and in part by the R-GIRO Research Fund from Ritsumeikan University

References

1. Wakamiya, M., Furukawa, A., Kanasaki, S., Murata, K.: Assessment of small bowel motility function with cine MRI using balanced steady-state free precession sequence. J. Magn. Reson. Imaging **33**(5), 1235–1240 (2011). doi:10.1002/jmri.22529
2. Wu, X., Chen, Y.W., Xi, Q., Zhang, S., Furukawa, A., Kanasaki, S., Wakamiya, M., Murata, K.: A semi-automated detection for motility of small bowel with MRI sequence. IEICE Tech. Rep. Image Eng. **109**(63), 101–105 (2009)
3. Furukawa, A., Chen, Y.W., Kanasaki, S., Wakamiya, M., Murakami, Y., Sato, S., Yoshimura, M., Tateyama, T., Taniguchi, A.: MR imaging in gastrointestinal tracts: static and dynamic assessment. In: Proceedings of ISSDM2012, Taiwan, pp. 441–445 (2012)
4. Taniguchi, A., Furukawa, A., Kanasaki, S., Tateyama, T., Chen, Y.W.: Automated assessment of small bowel motility function based on three-dimensional zero-mean normalized cross correlation. In: Proceedings of CISP-BMEI 2013, Hangzhou, China, pp. 802–805 (2013). doi:10.1109/BMEI.2013.6747050
5. Otsu, N.: A threshold selection method from gray-level histogram. IEEE Trans. Syst. Man. Cybern. **9**, 62–66 (1979). doi:10.1109/TSMC.1979.4310076
6. Beucher, S., Lantuéjoul, C.: Use of watersheds in contour detection. In: Proceedings International Workshop on Image Processing, Real-Time Edge and Motion Detection ∼ Estimation, Rennes (1979)
7. Caselles, V., Kimmel, R., Sapiro, G.: Geodestic active contours. J. Comput. Vis. **22**, 61 (1997). doi:10.1023/A:1007979827043
8. Achanta, R., Shaji, A., Smith, K., Lucchi, A., Fua, P., Süsstunk, S.: SLIC superpixels compared to state-of-the-art superpixels methods. IEEE Trans. Pattern Anal. Mach. Intell. **34**, 2274–2282 (2012). doi:10.1109/TPAMI.2012.120

Healthcare Support System

Template-Matching-Based Tracking of Cervical Spines in Videofluorography During Swallowing

Kojiro Mekata[1(✉)], Hotaka Takizawa[1], Tomoyuki Takigawa[2], Kazukiyo Toda[3], Yasuo Ito[3], and Hiroyuki Kudo[1,4]

[1] Department of Computer Science, Graduate School of Systems and Information Engineering, University of Tsukuba, Ibaraki, Japan
k.mekata@mibel.cs.tsukuba.ac.jp
[2] Department of Orthopaedic Surgery, Okayama University Hospital, Okayama, Japan
[3] Department of Orthopaedic Surgery, Kobe Red Cross Hospital, Hyogo, Japan
[4] JST-ERATO Quantum-Beam Phase Imaging Project, Sendai, Japan

Abstract. In our previous study, we focused on the motion of cervical spines in video-fluorography (VF), and we revealed that the physiological lordosises were reduced during normal swallowing. In the study, cervical spines were required to be manually extracted in all the frames of VF. Therefore the study had a difficulty in performing the statistical analysis based on a large amount of patient data. The present study proposes an automatic tracking method of cervical spines in VF by use of the two-dimensional template matching technique. In the method, cervical spines can be extracted automatically, though templates should be set manually only in the first frame of VF. The automatic tracking method was applied to cervical spines, from C1 to C6, in actual VF of ten cases who were planned to undergo posterior cervical spinal fusion. The cervical spines were able to be tracked at the accuracy of more than 77% in the Jaccard index. The experimental results demonstrated that the proposed method was able to analyze the motion of cervical spines in VF during swallowing.

Keywords: Swallowing · Cervical spines · Tracking · Two-dimensional template matching · Videofluorography

1 Introduction

Dysphagia makes swallowing difficult, and causes several diseases such as aspiration pneumonia. Many patients suffer from the dysphagia, and therefore it is necessary to analyze the mechanism of dysphagia for accurate diagnosis and adequate treatment. There have been many studies to determine the causes of dysphagia [1]. Several studies have examined the relationship between swallowing and cervical spines in several articles. Cervical spine disorders, such as cervical osteophytes and cervical exostosis, may cause dysphagia [2]. The incidence of dysphagia after anterior cervical surgery has been reported to be high. The sensation of impaired swallowing persists in patients undergoing anterior cervical discectomy and fusion even at six months after the surgery [3].

© Springer International Publishing AG 2018
Y.-W. Chen et al. (eds.), *Innovation in Medicine and Healthcare 2017*, Smart Innovation, Systems and Technologies 71, DOI 10.1007/978-3-319-59397-5_20

Posterior cervical fusion surgery without the anterior approach has been also reported to cause dysphagia two weeks after the surgery [4].

We have investigated a relation between swallowing and the motion of cervical spines. The study [5] demonstrated that the physiological lordosis of cervical spines were reduced during swallowing in videofluorography (VF) of thirty nine normal subjects. The study [6] demonstrated that the rotation of from C1 to C3 were restricted when subjects wore cervical orthoses, and that the restriction affected hyoid anterosuperior movement, epiglottis inversion, and pharyngoesophageal junction opening. These studies implied that there were relations between swallowing and the motion of cervical spines. One of problems of these studies was that cervical regions should be extracted manually from VF, and the manual extraction prevented investigation based on more patients. Normal swallowing motion synchronize hyoid bone elevation, and soft palate elevation, epiglottis inversion, laryngeal vestibule closure, and pharyngoesophageal junction opening. But, dysphagia can not synchronize those. Therefore, data analysis in limited frames is insufficient, and it is necessary to continuously analyze multiple cases over a long time. In the previous study, we proposed an automatic tracking method of cervical vertebral bodies based on template matching technique. This method was applied to the VF of three patients to undergo cervical spine fusion [7]. In this study, we proposed an automatic tracking method of cervical spines based on template matching technique. In this method, cervical spines can be extracted automatically, though templates should be set manually only in the first frame of VF. This method was applied to the VF of ten patients to undergo cervical spine fusion, and experimental results were shown.

2   Method

We used the VF data of subjects according to the instructions on the manual made by the Japanese society of dysphagia rehabilitation [8]. CUREVISTA (Hitachi Medical Corporation) and DMCAT-2000HL (Panasonic) were used to acquire and record the VF, respectively. The subjects were instructed to sit on seats and to set their necks perpendicular to the floor as much as possible (Fig. 1). The 10 ml of 40% diluted barium sulfate was used as contrast agents to allow us to confirm swallowing dynamics in the VF. Such contrast agents were required to be used by the manual, and therefore we did not obtain VF without contrast agents. The frame rates were set to be 30 per second. Shooting time was at most one minute. The total radiation dosage per VF examination was estimated at 0.5 mGy. A reference marker with a size of one cm was placed when capturing images. Image resolution was 0.5 mm/pixel.

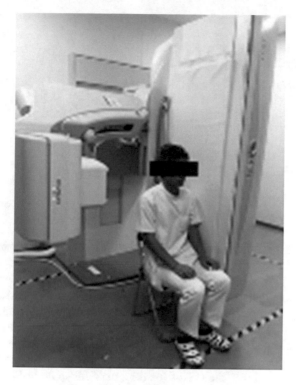

Fig. 1. Videofluorography system.

Figure 2 was the pharyngeal frame of VF. The overview of our tracking method was shown in Fig. 3. First, a boundary box (template) was manually set to each of cervical spines, from C1 to C6, in the first frame (t = t).

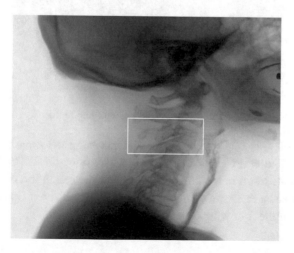

Fig. 2. A template (white square) for the third cervical spines in the pharyngeal frame of VF.

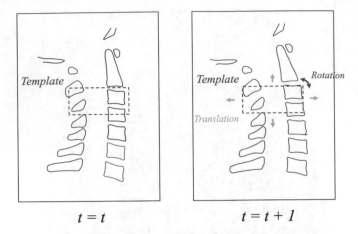

Fig. 3. Tracking process based on the template matching technique.

Figure 4 was the enlarged view of the template in Fig. 2. In order to evaluate the accuracy of the tracking, a clinical expert manually created the ground truth (GT) of the cervical spines (which were composed of cervical spinous processes and cervical vertebral bodies) from C1 to C6 in the VF frames of the pharyngeal and oral phases. The pharyngeal phase was defined to be the time when the hyoid bones reached at the most upper-front positions, and the oral phase was defined to be one second before the pharyngeal phase. Figure 5 was the GT of the third cervical spine in Fig. 4. The cervical spine regions were filled and then dilated as shown in Fig. 6. The circular structuring element with a radius of two pixels was used for the dilation.

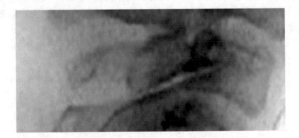

Fig. 4. Enlarged view of the template in Fig. 2.

In the next frame ($t = t + 1$), the template was set at the same position, and then translated in a range of $\pm U_{tmp}$ pixels in the x and y directions. At each position, the template was rotated in a range of $\pm \theta_{tmp}$ degrees at the resolution of $\Delta \theta_{tmp}$ degrees to create new templates. Each new template $T_{u,v,\theta}(x, y)$ and the next frame $F_{t+1}(x, y)$ were matched by use of the following Cosine similarity:

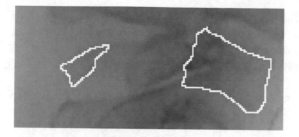

Fig. 5. Ground truth area of the third cervical spine in Fig. 4.

Fig. 6. Dilation result of the template in Fig. 4.

$$S(u, v, \theta) = \frac{\sum_{x,y} \left\{ T_{u,v,\theta}(x,y) \cdot F_{t+1}(x,y) \cdot mask(x,y) \right\}}{\sqrt{\sum_{x,y} \left(T_{u,v,\theta(x,y) \cdot mask(x,y)} \right)^2} \cdot \sqrt{\sum_{x,y} \left(F_{t+1}(x,y) \cdot mask(x,y) \right)^2}} \tag{1}$$

where (u, v) was the translation parameters and θ was the rotation parameter. $mask(x,y)$ was the value of the dilated template. The template with the maximum $s(u, v, \theta)$ was adopted as the optimal template. These processes were iterated.

3   Experiment

3.1   Experimental Conditions

We explained our research purposes, procedures, and voluntary participation to subjects undergoing posterior cervical spinal fusion. We obtained informed consent from ten subjects, who participated in the study. They were allowed to leave off the study at any time at their own will. The proposed method was applied to the VF of the subjects.

The GT at the oral phases were used as initial templates. The tracked templates were compared with the GT at the pharyngeal phases, and the accuracy was evaluated by use of the following Jaccard indexes:

$$JI = \frac{R_{t=1}^{TR} \cap R_{t=35}}{R_{t=1}^{TR} \cup R_{t=35}} \tag{2}$$

where $R_{t=1}^{TR}$ was the translated and rotated cervical spine region at the pharyngeal phase, and $R_{t=35}$ was the GT region at the oral phase. This value takes values from one to zero according to the overlap ratio of the two regions. The value takes one if they are overlapped completely, whereas zero if they have no overlap. In this study, the JI is used as an index value to evaluate the accuracy of tracking.

In this experiment, U_{tmp}, ΔU_{tmp}, θ_{tmp}, and $\Delta\theta_{tmp}$ were set to be 20 pixels, 1 pixels, 4° and 0.5°, respectively.

3.2   Results

The experimental results are shown in Table 1. The cervical spines can be tracked at more than 77%.

Table 1. Evaluation results of the cervical spines tracking for ten cases.

	C1	C2	C3	C4	C5	C6	Mean ± SD
Sample A	0.85	0.82	0.89	0.85	0.88	0.88	0.86 ± 0.03
Sample B	0.87	0.81	0.89	0.87	0.88	0.87	0.87 ± 0.03
Sample C	0.84	0.83	0.83	0.82	0.86	0.85	0.84 ± 0.02
Sample D	0.82	0.86	0.85	0.77	0.83	0.86	0.83 ± 0.04
Sample E	0.79	0.92	0.87	0.88	0.88	0.86	0.87 ± 0.05
Sample F	0.82	0.91	0.90	0.81	0.87	0.86	0.86 ± 0.05
Sample G	0.85	0.91	0.92	0.90	0.91	0.88	0.90 ± 0.03
Sample H	0.78	0.91	0.89	0.91	0.82	0.87	0.86 ± 0.06
Sample I	0.84	0.84	0.89	0.89	0.88	0.88	0.87 ± 0.03
Sample J	0.81	0.92	0.91	0.90	0.85	0.81	0.87 ± 0.05
Mean ± SD	0.83 ± 0.03	0.87 ± 0.05	0.88 ± 0.03	0.86 ± 0.05	0.87 ± 0.03	0.86 ± 0.02	

4   Discussion

Our final goal is to analyze the dynamics of swallowing in a large amount of VF, and to suggest adequate treatment to patients with dysphagia. We focused on the movement of cervical spines during swallowing, and therefore needed to track cervical spines in VF. We adopted two dimensional template matching technique. This technique is well known, but suitable for our purpose because the cervical spines move only two dimensionally and are not deformed. The main contribution of the paper is to be able to confirm that we can automatically track the cervical spines at the accuracy of more than 77%. This automatic tracking will allow us to statistically analyse the movement of cervical spines in a large amount of VF in the next step of our research.

In this method, templates should be manually set to target cervical spines in the first frames of VF as described above. In order to make our method fully automatic, such cervical spines should be recognized even in the first frames. We will consider to make use of several machine learning methods such as deep learning in the future.

A clinical expert visually confirmed that the tracking was good. In this experiment, the tracked regions were from C1 to C6. It is necessary to track the occipital bones and the seventh cervical spines. The maximum Jaccard value was 92%, whereas the minimum value was 77%. Tracking accuracy was not low, but it should be increased by, for example, reducing noises in VF beforehand.

In this study, we used the handwritten GT data without evaluating the accuracy of GT data itself. It is necessary to provide more reliable GT in the future.

5 Conclusion

In this paper, we proposed an automatic tracking method of cervical spines in VF during swallowing by use of a template matching technique. A clinical expert created templates for target cervical spines, from C1 to C6, in the first frame of VF, and the templates were matched with the subsequent frames to extract and track the cervical spines. The fidelity between the templates and the VF frames was evaluated by the Cosine similarity. This tracking method were applied to the VF data of ten cases, and the tracking accuracy was evaluated by Jaccard index between the tracked templates and GT data that were also created by the clinical export. The Jaccard indexes were from 77% to 92%. The experimental results demonstrated that the proposed method was promising as a means of tracking cervical spines in VF during swallowing.

One of our future works is to create more reliable GT data.

References

1. Logemann, J.A.: Evaluation and Treatment of Swallowing Disorders. College-Hill Press, London (1998)
2. Pimenta, A.P.: Dysphagia and cervical exostoses. Arq. Gastroenterol. **16**, 86–90 (1979)
3. Stachniak, J.B., Diebner, J.D., Brunk, E.S., et al.: Analysis of prevertebral soft-tissue swelling and dysphagia in multilevel anterior cervical discectomy and fusion with recombinant human bone morphogenetic protein-2 in patients at risk for pseudarthrosis. J. Neurosurg. Spine **14**, 244–249 (2011)
4. Radcliff, K.E., Koyonos, L., Clyde, C., et al.: What is the incidence of dysphagia after posterior cervical surgery? Spine **38**, 1082–1088 (2013)
5. Mekata, K., Takigawa, T., Matsubayashi, J., et al.: Cervical spine motion during swallowing. Eur. Spine J. **22**, 2558–2563 (2013)
6. Mekata, K., Takigawa, T., Matsubayashi, J., et al.: The Effect of the cervical orthosis on swallowing physiology and cervical spine motion during swallowing. Dysphagia **31**, 74–83 (2016)
7. Mekata, K., Takizawa, H., Matsubayashi, J., et al.: A preliminary study on template-matching-based tracking of cervical vertebral bodies in videofluorography during swallowing. Int. Forum Med. Imaging Asia (IFMIA) **2017**, 190–191 (2017)
8. The Japanese society of dysphagia rehabilitation. http://www.jsdr.or.jp/wp-content/uploads/file/doc/VF18-2-p166-186.pdf. Accessed 5 Jan 2016

Development of Learning Materials to Support Assisting-Skill Acquisition Using 3DCG

Kaoru Eto[1]([⊠]), Hiroshi Matsuda[1], Hiroshi Takase[1],
Atsuko Yamazaki[2], Hiroko Yoshida[3], Kiyomi Ito[4], Michie Ogiwara[5],
Kikuko Saeki[6], and Akihiko Shimizu[7]

[1] Faculty of Engineering, Nippon Institute of Technology, Miyashiro, Japan
{eto,hiroshi,takase}@nit.ac.jp
[2] College of Engineering, Shibaura Institute of Technology, Omiya, Japan
atsuko@sic.shibaura-it.ac.jp
[3] Itabashi Royal Home-Visit Nursing Station, Tokyo, Japan
itaroist@po.wol.ne.jp
[4] Higashi Honchou Home-Visit Nursing Station, Tokyo, Japan
kti-itou@kki.biglobe.ne.jp
[5] Jiyugaoka Home-Visit Nursing Station, Tokyo, Japan
jiyuugaoka.st@cc.wakwak.com
[6] Itabashi Central Nursing School, Tokyo, Japan
kil0ko26rin0823@ac.auone-net.jp
[7] Itabashi Chuo General Hospital, Tokyo, Japan
a.shimizoo@gmail.com

Abstract. This study developed learning materials to support the acquisition of skills for elderly care. Our aim is to establish a method that develops human resources for elderly care using ICT and develop learning materials for caregivers. We developed learning materials that increase the awareness of novice caregivers based on comparisons of their care with the care of experts by overlapping the movements of novices and experts in 3DCG movies. Evaluation and feedback from a nurse, a physical therapist, an orthopedist, and a nursing school teacher confirmed that 3DCG movies and overlapping techniques can provide useful educational benefits.

Keywords: 3DCG · Assisting-skill acquisition · Visualization · Overlapping movements

1 Introduction

This study developed learning materials to support the acquisition of the skills required for elderly care. In Japan, since the number of senior citizens over 75 will swell to more than eight million by 2025 [1], we must develop methods to cultivate human resources for elderly care. We established a method that develops human resources for elderly care using ICT and created learning materials for caregivers. We focused on increasing the awareness of novice caregivers of the differences between their own care and that of experts by overlapping the movements of novices and experts in 3DCG movies.

© Springer International Publishing AG 2018
Y.-W. Chen et al. (eds.), *Innovation in Medicine and Healthcare 2017*, Smart Innovation, Systems and Technologies 71, DOI 10.1007/978-3-319-59397-5_21

2   Previous Research

Images that provide educational benefits can be imparted for master nursing skills [2]. Such benefits have also increased nursing skills by making comments in videos based on nursing skills, leading to improvements [3]. However, a problem exists; movies and images provide excessive information, complicating the extraction of crucial information. On the other hand, 3DCG movies can easily depict care movements and critical information [4].

We developed a system that extracts, shares, and utilizes expert know-how in the care plan formulation process and clarified its effectiveness from the results of trials at a care site. In our system, the assessment results of the care recipients are visualized on circular charts, which are read by experts who calculate them based on a concept that reflects keywords that are arranged and visualized two-dimensionally [5, 6]. We are currently verifying the effectiveness of our system by visualizing the effects of foot care [7, 8]. Based on these research results, we believe that our system can acquire care skills using 3DCG and will be an effective tool for mastering senior-care skills.

3   Method of Learning Support

Based on the results of the above research, we used the following two strategies to support the acquisition of assisting skills:

(1) With video cameras, we shot scenes where novices and experts performed identical assisting skills and made 3DCG movies. The learning of novices is supported by a feature where a patient's entire body can be observed from many directions in 3DCG.
(2) We overlapped the movements of novices and experts in 3DCG movies and focused on increasing the awareness of novice caregivers of the differences between their care and that of experts by overlapping images. We shot assisting skills using body mechanics for our comparisons of novices and experts.

3.1   Features of 3D CG Movies

The features of the 3DCG images are shown below:

(1) 3DCG images can compare depths, which is impossible in two dimensions.
(2) They can observe the area behind an object that cannot be seen with 2D video.
(3) With 3DCG imaging, we can observe the patient's body from various angles while rotating the video and from angles that cannot be seen by video.
(4) They can be scrutinized more minutely by slowing down or stopping the moving images.

3.2   3DCG Imaging Method and Procedure

Next we describe our method and procedure of shooting the assistance movements to 3DCG imaging and overlapping images.

Shooting assisting skills
With a video camera, we shot scenes where novices and experts are performing the same assistive action. During shooting, the angle is critical for getting images that can be traced. The angles were set to capture the entire body and the movements of the assistants and the caregivers. In this study, a nursing student's assisting skills were captured as a novice image, and for the images of experts, we used the images of a nursing school teacher.

3DCG imaging
We used Miku Miku Dance (MMD) on the captured images to make 3DCG images. MMD is software that can trace the motion of a person or an object from movies taken with a video camera to an MMD model for 3DCG imaging.

In this study, we shot the assisting skills of novices and experts and traced 3DCG images that visualized the assisting skills. Figure 1 shows a 3D model that is traced on the background of the assisted action that was taken. We can instantly check the created motion as a moving image, trace the small movements, and express the differences in behavior between novices and experts in three dimensions.

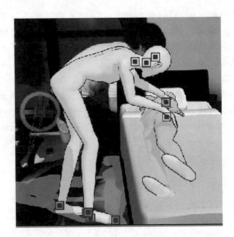

Fig. 1. Tracing with MMD

Overlapping 3DCG images
We overlapped the images of novices and experts as 3DCG images. From the original motion data created by MMD, we created 3D models by Blender, PMDEditor and overlapped the experts' motion data. To compare the novices and the experts, the 3D model represented the novices and caregivers as well as the experts and caregivers in different colors. Figure 2 is a novice 3DCG model. On the left is the assistant and the novice is on the right.

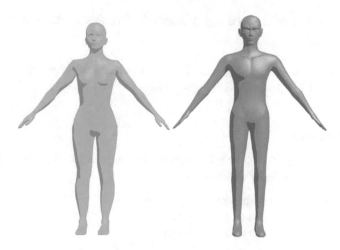

Fig. 2. 3DCG model of novices

Figure 3 is a 3DCG model of an expert. The figure on the left is the assistant, and the expert is on the right. Since the model is overlapped when the movie is played by unifying the starting positions of the models of the novice and the expert as well as the movement positions of each model of the caregivers, the four models performed individual actions. We captured the differences between the experts and novices.

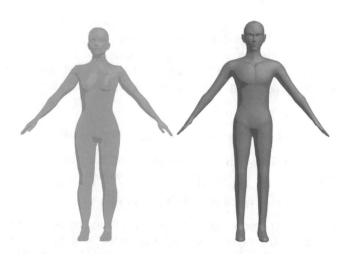

Fig. 3. 3DCG model of experts

4   Development of Learning Materials for Assisting-Skill Acquisition

For the development of teaching materials, we selected obligatory nursing skills and inserted narration into the 3DCG imaged videos.

4.1   Assisting Skills

The content of our developed learning material is comprised of the basic operations of the following three kinds of assisting skills:

(1) From a horizontal movement to a prone position;
(2) From a supine position to a sitting position;
(3) From a sitting position to a wheelchair.

In the assisting process, the operations include such basic daily human movements as eating, defecation/urination, attaching and detaching the clothes, and bathing.

4.2   Movie Editing

We used Windows Movie Maker for the 3DCG images created by MMD and edited the movies to reproduce the important points in slow motion.

4.3   Creation of Narration

We used VOICEROID, which is speech synthesis software, to make a 3DCG movie of assistive technologies and inserted operation explanations and points to be attended by the learner as narration based on the assisting procedure. For example, when the upper body starts to rise, the front of the hip joint is held with the hand that supports the knee to stabilize the sitting position so that both plantars are placed on the floor.

4.4   Learning Materials to Support Assisting-Skill Acquisition

Each of the above assisting skills can be viewed as learning material for a novice, an expert, and overlapping motions of a novice and an expert.

(1) **From a horizontal movement to a supine position.**

Figure 4 shows the novice's operation, and Fig. 5 shows an expert's operation for a horizontal movement. The expert is clearly putting her arms deeper underneath the patient than the novice. The expert is also lowering her hips more than the novice. Figure 6 shows overlapping images of the expert and the novice. By overlapping them, their subtle differences become more conspicuous, and the novice's flaws are highlighted.

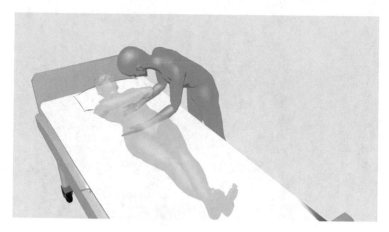

Fig. 4. Novice's horizontal movement

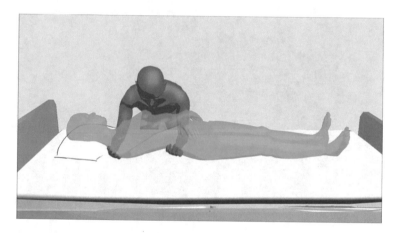

Fig. 5. Expert's horizontal movement

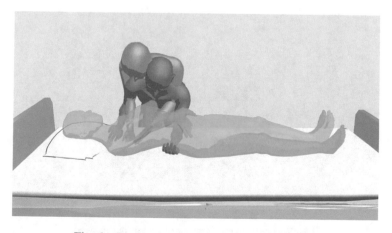

Fig. 6. Overlapping images of expert and novice

Fig. 7. Sitting position operation of novice

Fig. 8. Sitting position operation of expert

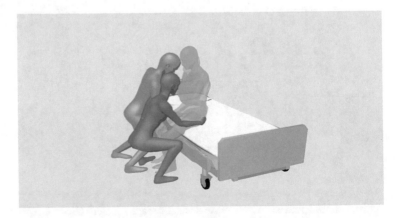

Fig. 9. Overlapping images of expert and novice

(2) **From supine to a sitting position.**

Figure 7 shows the novice's operation, and Fig. 8 shows the expert's operation of the patient's sitting position. In Fig. 6, the novice is starting the movement without folding both of the patient's arms on her chest. On the other hand, in Fig. 8, the expert first folded the patient's arm to reduce the frictional resistance. In Fig. 9, the expert steadily lowered her center of gravity to reduce the burden on her hips.

(3) **From a sitting position to a wheelchair.**

In assisting skills, the most technical differences appear between experts and novices in the movement that transfers a patient to a wheelchair.

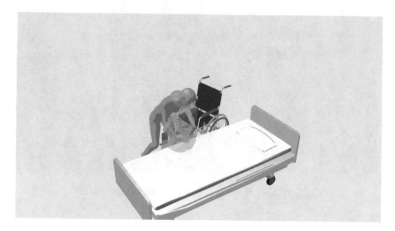

Fig. 10. Being transferred to a wheelchair by novice

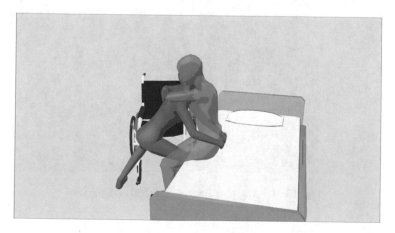

Fig. 11. Being transferred to a wheelchair by expert

Comparing the arm positions of the novice and expert with the height of their hips (Figs. 10 and 11), the novice's arm is higher than the expert's, and the hips of the novice are also higher than those of the expert. The problem with the position of the feet is obvious. In Figs. 12 and 13, when comparing the height of the hips of the beginners and the experts and the position of the patient's arm, the difference in the hip's height is clear. The position of the novice's arm is higher than the expert's arm, and the hips of the novice are higher than those of the expert. The problem with the difference in the feet's position is again obvious.

Fig. 12. Overlapping images of expert and novice

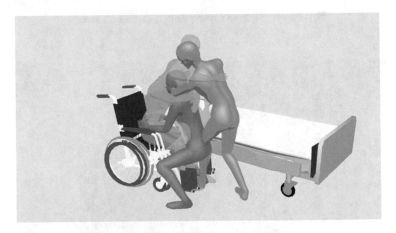

Fig. 13. Overlapping images of expert and novice

5   Discussion and Conclusion

In this study, we developed learning materials to support assisting-skill acquisition using 3DCG movies for novice caregivers. We examined our learning materials with 17 persons (a nursing school teacher, eight nurses, five physical therapists, an orthopedist, and two caregivers) and obtained the following results:

(1)   Since comparing overlapping images of experts and novices shows that good and bad examples can be shown simultaneously, images of accurate assisting skills of all of the assisting movements can be easily grasped.
(2)   Since the assisting skills can be seen from above, such images are clearly visible by checking places that are basically hidden or out of sight.
(3)   When novices cannot remember information/techniques from previous classes and lectures, they can repeatedly review 3DCG movies and reduce their re-training or review time.
(4)   We demonstrated the educational impact of technology and its power.

These factors reflect the fact that the features of 3DCG images are exploited as learning materials. We also highlighted the influence of overlapping the 3DCG images of experts and novices. From these evaluations, we confirmed that positive educational impact can be achieved by overlapping 3DCG visualization and its images. In the next study, we will develop evaluation criteria to improve 3DCG movies and the viewing system.

References

1. http://www.mhlw.go.jp/english/policy/care-welfare/care-welfare-elderly/index.html
2. Soga, H., Yoshikawa, H., Shioduki, T., Adachi, M., Morikawa, S.: Efforts to introduce 3D stereoscopic display teaching materials to learning of morphological functional studies. J. Nurs. **12**(1), 65–68 (2014). (in Japanese, Shiga University of Medical Science)
3. Majima, Y., Maekawa, Y., Shimada, S., Izumi, T.: Improvement program of nursing skill movie manual by movie-based comment system, IEICE Tech. Rep. ET2013-9 (2013). (in Japanese)
4. Sato, K., Kaiga, T., Watabe, S.: Development of application model for dance education using motion capture. IPSJ SIG Tech. Rep. **2009-CH-82**(6) (2009). (in Japanese)
5. Eto, K., Matsui, T., Kabasawa, Y.: Development of know-how information sharing system in care planning processes-mapping new care plan into two-dimension document space. In: LNCS, vol., 4252, pp. 977–984. Springer, Heidelberg (2006)
6. Eto, K., Matsui, T., Kabasawa, Y.: Educational effect of externalization of know-how information for care planning processes. In: Advanced Learning, pp. 423–443. InTech (2009)
7. Eto, K., Yamazaki, A., Yonekura, K., Mukuda, M., Kabasawa, Y., Yoshida, H., Ito, K., Ogiwara, M.: A preliminary examination of effect of massage and aroma oil massage in foot care nursing. Procedia Comput. Sci. **60**, 1524–1531 (2015)
8. Eto, K., Yamazaki, A., Yonekura, K., Mukuda, M., Kabasawa, Y., Yoshida, H., Ito, K., Ogiwara, M.: Visualization of effects of aroma oil massages using NIRS. Procedia Comput. Sci. **96**, 1535–1542 (2016)

Feature Selection and Machine Learning Based Multilevel Stress Detection from ECG Signals

Isabelle Bichindaritz[✉], Cassie Breen, Ekaterina Cole, Neha Keshan, and Pat Parimi

Computer Science Department,
State University of New York at Oswego, Oswego, USA
{ibichind, cbreen, ecole3, neha.keshan,
patanjali.parimi}@oswego.edu

Abstract. Physiological sensor analytics aims at monitoring health as the availability of sensor-enabled portable, wearable, and implantable devices become ubiquitous in the growing Internet of Things (IoT). Physiological multi-sensor studies have been conducted previously to detect stress. In this study, we focus on electrocardiography (ECG) monitoring that can now be performed with minimally invasive wearable patches and sensors, to develop an efficient and robust mechanism for accurate stress identification, for example in automobile drivers. A unique aspect of our research is personalized individual stress analysis including three stress levels: low, medium and high. Using machine learning algorithms from the ECG signals alone, our system achieves up to 100% accuracy and area under ROC curve of 1 depending on the experimental setting in detecting three classes of stress using feature selection from a combination of fiducial points and multiscale entropy as a fine-grained indicator of stress level.

Keywords: Machine learning · Sensors · Data mining · Stress medicine · ECG

1 Introduction

According to the National Highway Traffic Safety Administration (NHTSA) and the Virginia Tech Transportation Institute (VTTI), lack of attention while driving is found to be the leading cause of automobile accidents in the US in 80% of all crashes and 65% of all near-crashes. These data clearly show that improved attention and close monitoring of drivers' conditions could help increase their safety. Driving in stressful environments such as city or highway prompts for drivers' heightened attention and is also correlated with higher risk of accidents because prolonged stress decreases one's ability to be attentive.

With the availability of portable wearable and implantable devices in the growing Internet of Things (IoT), physiological sensor data analytics will lead to improved health care monitoring [1] and preventive care [2]. Although physiological multi-sensor studies have been conducted with some success to detect stress based on such measures as heart rate variability, skin conductance, respiration rate, electromyogram (EMG), body

© Springer International Publishing AG 2018
Y.-W. Chen et al. (eds.), *Innovation in Medicine and Healthcare 2017*, Smart Innovation, Systems and Technologies 71, DOI 10.1007/978-3-319-59397-5_22

temperature, blood pressure, and electro-encephalogram (EEG). Electrocardiography (ECG) has often been discarded by these studies due to the constraints of the measurements requiring 16 leads and the possible imperfections of the resulting signals, which can fail to detect some heart beats. However, ECG signals are highly valued for the precision of their R-peak detection, leading to excellent heart rate rhythm measurement after preprocessing the signal for missed beats [3]. In addition, ECG monitoring can now be performed with minimally invasive wearable patches and other sensors, which makes stress detection based on them an interesting field of study [4].

In this study, we apply machine learning methods and algorithms to detect stress from ECG signals in subjects under different levels of environmental stress caused by driving conditions. We find that stress levels can be successfully detected from ECG signals alone; with random tree classifier allowing for identification of the three classes of stress, low, medium and high, with up to 100% accuracy depending on the experimental setting, which is a significant improvement on a prior study on the same data set [5]. In particular, classification accuracy was improved by 10% in cross-validation with Multilayer Perceptron.

2 Background

The ECG is one of the simplest and oldest cardiac monitors available and yet it can provide a wealth of useful information. ECG represents the electrical activity of the heart muscle as it changes with time [6]. Like other muscles, the cardiac muscle contracts in response to electrical *depolarization* of the muscle cells. It is the sum of this electrical activity, when amplified and recorded for just a few seconds that is known as an ECG.

Important waveforms of an ECG are marked as P, Q, R, S and T (see Fig. 1) and represent the changes in electrical potential as the heart contracts and relaxes. Points P, Q, R, S and T are called fiducial points. Depolarization of the ventricles results in usually the largest part of the ECG signal (because of the greater muscle mass in the ventricles) and this is known as the **QRS complex** [7].

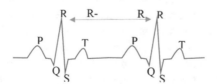

Fig. 1. An ECG signal depicting the RR interval

- The Q wave is the first initial downward or 'negative' deflection.
- The R wave is the next upward deflection (provided it crosses the isoelectric line and becomes 'positive')
- The S wave is the next deflection downwards, provided it crosses the isoelectric line to become briefly negative before returning to the isoelectric baseline.

3 Materials and Methods

3.1 Data

The ECG signals of stress used in this study were obtained from MIT-BIH PhysioNet Multi-parameter Database [8]. These data sets are part of the experiment conducted by Healey and Picard [3] and has data from 17 participating drivers and eight types of raw data – time stamp, ECG, electromyogram (EMG), foot galvanic skin response (GSR), hand GSR, intermittent heart rate (IHR), marker, and respiration – all acquired from different wearable sensors. During the experiment conducted by Healy and Picard, the drivers drove in Greater Boston area from MIT's East Garage to River Street Bridge and back through three cities and two highways. The initial rest and final rest states, as well as stress during driving were measured.

The data sets were segmented into three stress levels – low stress (initial rest and final rest), moderate stress (highway), and high stress (cities) – assuming that the stress acquired by subjects is solely based on traffic conditions and for no other reason. The signal classification was carried out by considering the variation in the ECG signals of the three states, low, medium and high, in an individual.

The time durations for each segment – rest, highway, and city - given by Akbas were used to distinguish between the rest, highway and city time periods [9]. The segmentation mark of different driving periods was not clear in seven of the data sets as was also found by Akbas [9]. Consequently, only 10 drivers' data sets were used for this study. Using the methods available from Physionet [8], an annotation was performed on each data set separately and annotated files were obtained for each driving period of the ten drivers.

3.2 Feature Extraction

Feature extraction was performed to extract 14 different fiducial points (P, Q, R, S) interval features, averaged over the time intervals (see Table 1, left column) from the annotated ECG signals annotations using NetBeans Java platform (see Fig. 2) to produce the required file for classification in Waikato Environment for Knowledge Analysis (Weka) [10]. We have considered all possible signal attributes and their relations (Table 1) in feature extraction to carry out a thorough analysis. We demonstrated in a previous paper [5] that near-perfect classification could be achieved with these 14 features alone – and even a subset of these – for two stress levels. However, results for three stress levels were not convincing, which prompted us to add multiscale entropy to this original set of features.

Therefore, we added variance for these intervals (12 new features), which did not improve the results much. We then performed multiscale entropy analysis [11] of the annotation files. Intuitively, the entropy of a signal measures the amount of disorder and complexity present in this signal. Pathological states have been found to be associated with decreased complexity in signals, and lower multiscale entropy. Examples of pathologies include aging and chronic heart failure. Multiscale entropy consists in extracting from a time series entropy measures associated with several

Table 1. Extracted features from the ECG signals

Fiducial features		Entropy features	
Average/Var QRS interval	Average/Var QRS difference	A1 to A20 entropy	A1 to A20 entropy diff.
Average/Var RR interval	Average/Var RR difference		Average entropy diff.
Average/Var QQ interval	Average/Var QQ difference	Entropy variance	Entropy variance diff.
Average/Var SS interval	Average/Var SS difference	Entropy slope 1	Entropy slope 1 diff.
Average/Var QR interval	Average/Var QR difference	Entropy slope 2	Entropy slope 2 diff.
Average/Var RS interval	Average/Var RS difference		
Average beats	Average difference beats		

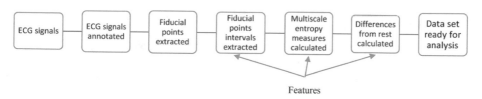

Features

Fig. 2. Feature extraction algorithm flowchart

scales in the signal, according to windows of varying level of granularity, starting from the whole signal and progressively dividing it by a scale factor. The algorithm proceeds in two steps. First for each scale i, a time series is generated by averaging the samples in windows of length i. Then the entropy is calculated for each coarse-grained time series by the conditional probability quantifying the likelihood that if two sets of simultaneous data points of a given length have distance <r, then two sets of simultaneous data points of the given length plus 1 also have distance <r [11].

With a scale factor of 20, we obtain 20 entropy measures (see Table 2). We can also calculate the average entropy and its variance, as well as the slopes at the beginning of its curve and at its end (see Table 2).

In addition, differences in fiducial point characteristics and multiscale entropy measurements between rest and stress states were recorded (Fig. 2 and Table 2) since the differences from the baseline may be important (as we found in a previous study [5].

Table 2. Accuracy percentage of three classes with 74 attributes

Accuracy (%)	Classification methods					
	Training set	Leave 1 out	2 Folds	10 Folds	75% Split	90% Split
Naïve Bayes	71.64	53.73	50.75	53.73	58.82	85.71
BayesNet	73.13	67.16	62.69	61.19	**70.59**	85.71
Logistic	**100**	38.81	53.73	47.76	52.94	57.14
Multi. Perceptron	98.51	50.74	62.69	56.72	58.82	85.71
SMO	85.07	59.70	53.73	58.21	58.82	57.14
IB1	**100**	41.79	47.76	38.81	52.94	85.71
IBK	**100**	41.79	47.76	38.81	52.94	85.71
KStar	**100**	53.73	50.75	53.73	58.82	**100**
ZeroR	44.78	44.78	44.78	44.78	29.41	41.86
J48	94.03	**68.66**	59.70	**62.69**	64.71	71.43
Random Forest	**100**	61.19	**65.67**	**62.69**	58.82	71.43
Random Tree	**100**	46.27	59.70	56.72	58.82	71.43

3.3 Classification

Predicting the level of stress from these ECG signals is a typical classification task in data mining. Three classes are available for classification purposes. Class '0' represents low stress (rest state), class '1' moderate stress (highway driving) and class '2' represents high stress (city driving).

Weka offers many classifiers out of which 12 algorithms from varied types were selected for classification to perform their comparative study [10] (see Table 2).

3.4 Assessment

The following six different test and experimental settings were applied on the ECG data sets: Training Set, Leave One Out Cross Validation (LOOC), 2-Folds Cross Validation, 10-Folds Cross Validation, 75% Split, and 90% Split.

The data set analyzed was small since it contained only 68 instances, obtained from the 70 potential instances for 10 drivers and 7 driving intervals for each [9]. Two of these driving intervals could not be analyzed to produce an annotation file with Physionet's annotators. From these 68 annotated signals, we have removed the data of drive05 highway1 as it is an obvious outlier with an average number of beats per minute of 29 – a highly unlikely figure. Other studies also reported this outlier and removed it. Therefore, the experiments presented below were conducted using the remaining data set of 67 instances with 74 different extracted attributes.

The experiments were conducted on 3 classes of stress – '0' for low stress, '1' for moderate stress, and '2' for high stress. The features extracted consisted in 14 fiducial measures and 48 entropy measures. Each time, classification was performed on all the features because the classification methods used were capable of selecting best features.

74 features were considered (see Table 2):

- 6 average interval durations (QRS, RR, QQ, SS, QR, RS).
- 1 average number of beats per minute.
- 6 average interval durations differences between initial rest and current state (QRS, RR, QQ, SS, QR, RS).
- 1 average number of beats per minute difference between rest and current state.
- 6 variance of interval durations (QRS, RR, QQ, SS, QR, RS).
- 6 variance of interval durations differences between initial rest and current state (QRS, RR, QQ, SS, QR, RS).
- 24 entropy measures (A1 to A20, average, variance, slope between A1 and A2, slope between A19 and A20).
- 24 differences in entropy measures between initial rest and current state (A1 to A20, average, variance, slope between A1 and A2, slope between A19 and A20).

4   Results

Accuracy, sensitivity (true positive rate), specificity (true negative rate), and area under ROC curve (AUC) were selected as performance measures. Accuracy was rounded off to two decimal places and the sensitivity, specificity, and AUC were rounded off to three decimal digits. Bold values in the tables represent the best results per column.

4.1   Results on 3 Classes and 74 Features

The results presented here are the classification accuracy percentage (Table 2), sensitivity/specificity, and AUC (Table 3) on the data set with three classes, all 74 attributes, and 67 instances.

4.2   Results on 3 Classes and 10 Selected Features

The results presented here are the classification accuracy percentage (Table 4), sensitivity/specificity, and AUC (Table 5) on the data set with three classes and 10 selected attributes chosen by automated feature selection in Weka (SVMAttributeEval with Ranker method).

4.3   Results Interpretation

The results on 3 classes and 74 features show that the highest accuracy was obtained for J48 (decision tree) on LOOC (68.66%) and Random Forest and J48 on 10-fold cross validation (62.69%). However, with a 90% split, which is acceptable for such a small data set, accuracy reaches 100% for KStar (Table 3). If considering the AUC, Random Forest reaches .832 in 10-fold cross validation and 1 with a 90% split (Table 3).

Table 3. Sensitivity/Specificity and AUC of three classes with 74 attributes

Sens/Spec AUC	Classification Methods					
	Training set	Leave 1 out	2 Folds	10 Folds	75% Split	90% Split
Naïve Bayes	.716/ .877/.891	.537/ .782/.646	.507/ .775/.611	.537/ .788/.658	.588/786/ .649	.857/ .976/1
BayesNet	.731/ .802/.844	.672/ .766/.72	.627/ .768/.741	.612/.75/ .745	**.706/ .801/.795**	.857/ .893/.839
Logistic	**1/1/1**	.388/ .683/.587	.537/ .791/.7	.478/ .735/.616	.529/ .768/.662	.571/ .679/.667
Multi. Perceptron	.985/ .988/.991	.507/ .747/.718	.627/ .812/.741	.567/ .757/.724	.588/ .786/.718	**.857/ .893/.976**
SMO	.851/ .918/.898	.597/ .782/.749	.537/ .766/.684	.582/ .784/.723	.588/ .786/.716	.571/ .845/.833
IB1	**1/1/1**	.418/ .673/.546	.476/ .722/.6	.388/ .656/.522	.529/ .761/.645	.857/ .893/.875
IBK	**1/1/1**	.418/ .673/.546	.476/ .722/.596	.388/ .656/.539	.529/ .761/.645	.857/ .893/.875
KStar	**1/1/1**	.537/ .748/.717	.507/ .755/.706	.537/ .748/.726	.588/ .786/.736	1/1/1
ZeroR	.448/ .552/0.5	.448/ .552/0.5	.448/ .552/0.5	.448/ .552/.461	.294/ .706/.5	.429/ .571/.5
J48	.94/.971/ .98	.687/ .836/.674	.597/.8/ .679	.627/ .794/.687	.647/ .789/.669	.714/ .869/.72
Random Forest	**1/1/1**	**.612/ .755/.818**	**.657/ .803/.799**	**.657/ .761/.832**	.588/ .758/.751	.714/ .869/.929
Random Tree	**1/1/1**	.463/ .723/.593	.597/ .808/.703	.567/ .788/.678	.588/ .833/.71	.714/ .869/.792

The results on 3 classes and 10 automatically selected features (see Table 4) show that the highest accuracy was obtained for J48 on LOOC (73.13%) and Multilayer Perceptron (MLP) on 10-fold cross validation (80.60%). However, with a 90% split, which is acceptable for such a small data set, accuracy reaches 100% for MLP and Logistic (Table 4). Considering the AUC, MLP ranks higher in most categories, including 10-fold cross-validation with .875 and 90% split with 1, also reached by Logistic (Table 5).

Overall, results have significantly improved using feature selection, with or without cross validation. However Random Forest, which performed best on 74 features, due to its ability to discriminate between features, was overtaken by MLP after feature selection.

In our previous studies, the results on 3 classes and 1 feature showed that the highest accuracy was obtained for MLP (neural network) on LOOC (68.66%) and 10-fold cross validation (70.15%), with J48 almost as accurate [5]. By adding variance and multiscale entropy, accuracy has improved by 10% with MLP in cross validation and by 14% in 90% split, which are significant improvements.

Table 4. Accuracy percentage of three classes with 10 selected attributes

Accuracy (%)	Classification Methods					
	Training set	Leave 1 out	2 Folds	10 Folds	75% Split	90% Split
Naïve Bayes	73.13	56.72	58.21	58.21	58.82	85.71
BayesNet	68.66	67.16	61.20	65.67	70.59	85.71
Logistic	86.57	71.64	56.72	65.67	41.18	**100**
Multi. Perceptron	94.03	71.64	**73.13**	**80.60**	**76.47**	**100**
SMO	77.61	67.16	62.69	70.15	58.82	85.71
IB1	100	64.18	64.18	65.67	64.71	71.43
IBK	100	64.18	64.18	65.67	64.71	71.43
KStar	100	53.73	50.75	55.22	58.82	85.71
ZeroR	44.78	44.78	44.78	44.78	29.41	42.86
J48	86.57	**73.13**	65.67	67.16	**76.47**	85.71
Random Forest	100	70.15	64.18	68.66	64.71	85.71
Random Tree	100	67.16	61.19	61.19	58.82	85.71

The best overall results were obtained for the features automatically selected by SVMAttributeEval (see Table 6), which combine the difference between rest and stress for each instance as well as some features independent from the rest state. Selected features include difference of heart rate from rest state, difference of variance in RR interval from rest state, difference of variance in QQ interval from rest state, variance in SS interval, heart rate, difference of variance in SS interval from rest state, difference of variance in entropy A10 and in entropy A7 from rest state, variance in entropy A4, and variance in entropy A13. We see that the addition of multiscale entropy has significantly improved the classification performance with four contributing features.

5   Discussion

The particular data set used for the present study was produced by Healey and Picard as part of Healey's PhD work [3]. These authors find a predictive accuracy of 97.4% with LOOC for high stress and 94.7% for moderate stress based on data extracted from EMG, respiration, instantaneous heart rate (extracted from ECG), and GSR as well as additional features. The focus of this research was in generalized identification of stress states using signal fusion of multiple sensors, but it is not tailored for an individual for whom the stress is classified into three states. The unique aspect of our research is personalized individual stress analysis using ECG data alone. In addition, the data they used is not exactly the same so that results are not completely comparable. Yet our results to detect high stress are comparable to Healey and Picard [3].

Akbas calculated the differences between the 3 stress levels of averaged feature values extracted from instantaneous heart rate, EMG, hand GSR, foot GSR, instantaneous respiration rate, and average number of contractions per minute [9]. This author

Table 5. Sensitivity/Specificity and AUC of three classes with 10 selected attributes

Sens/Spec/AUC	Classification methods					
	Training set	Leave 1 out	2 Folds	10 Folds	75% Split	90% Split
Naïve Bayes	.731/ .878/.88	.567/ .783/.77	.582/ .773/ .753	.582/ ,788/ .779	.588/ .752/ .835	.857/ .893/.976
BayesNet	.687/ .765/ .767	.672/ .773/ .581	.612/ .756/ .712	.657/ .768/ .713	.706/ .794/ .803	.857/ .893/.839
Logistic	.866/ .929/ .929	.716/ .834/ .830	.567/ .816/ .709	.657/ .818/ .834	.412/ .746/ .702	1/1/**1**
Multi. Perceptron	.94/.958/ .956	.716/ .827/ **.847**	.731/ .883/ **.815**	.806/ .899/ **.875**	.765/ .908/ **.932**	1/1/**1**
SMO	.776/ .883/ .853	.672/ .831/ .771	.627/.8/ .744	.701/ .848/ .795	.588/ .758/ .718	.857/ .893/.887
IB1	1/1/1	.642/ .779/ .711	.642/ .81/.726	.657/ .791/ .724	.647/ .832/ .739	.714/ .786/.75
IBK	1/1/1	.642/ .779/ .711	.642/ .81/.71	.657/ .791/ .765	.647/ .832/ .739	.714/ .786/.75
KStar	1/1/1	.537/ .753/ .689	.507/ .758/ .634	.552/ .752/ .693	.588/ .786/ .712	.857/ .893/**1**
ZeroR	.448/ .552/.5	.448/ .552/0	.448/ .552/ .492	.448/ .552/ .461	.294/ .706/.5	.429/ .571/.5
J48	.866/ .922/ .919	.731/ .851/ .711	.657/ .823/ .723	.672/ .822/ .731	.765/ .819/ .768	.857/ .893/.851
Random Forest	1/1/1	.701/ .815/ .827	.642/ .798/ .814	.687/ .815/ .842	.647/ .782/ .812	.857/893/ .893
Random Tree	1/1/1	.672/ .816/ .744	.612/ .812/ .712	.612/ .768/.69	.588/ .764/ .676	.857/ .893/.875

Table 6. SVMAttributeEval 10 selected features

1. avgDiffBeats	2. varDiffRR	3. varDiffQQ	4. varSS	5. avgBeats
6. varDiffSS	7. varDiffEntA10	8. varDiffEntA7	9. varEntA4	10. varEntA13

found these averages to be significantly different between the three levels of stress. However, no classification was performed.

Deng et al. extracted features from this data set based on principal component analysis (PCA) and determined that 5 features were best representative of this data set – foot GSR duration, hand GSR duration, hand GSR arca, foot GSR area, and foot GSR frequency [12]. These results are not really comparable to ours since we are using only ECG. However, these authors are also classifying the signals using machine learning algorithms and report best average rate of 75.38% on all features with NaiveBayes and 78.46% accuracy on the 5 selected features with SVM, using LOOCV. These results are not exactly comparable to ours because the authors removed two additional samples, which can alter results on such a small data set. However, we concur that feature selection improves classification accuracy over not selecting features.

Deng et al. pursued their research by combining feature selection with signal selection, reducing the number of signals used to 2 [13] in their preceding study. They selected 5 features based on C4.5 and 2 sensors. With 10-fold CV (averaged 6 times), they obtained accuracy of 74.5% with SVM on all features and 85.46% with C4.5 on 5 features. However, they used only 65 data samples. The same authors published another paper applying combinatorial fusion to the same task [14] with comparable results. Since the data set used is not the same, their results are not comparable to ours.

Singh and Queyam also combined all sensors for the classification task using neural networks. They reported good results of over 80% on 6 out of 10 drives [15]. They reported on selecting features as being more correlated with driving conditions, and they found that mean heart rate and mean hand GSR were the most correlated [16]. However, they did not use entropy measurements so that their results are not clearly comparable to ours because they used multiple sensors.

Avki et al. reported also on correlations between features and stress level [17]. They found that the variance in the signals measurements is the most correlated to stress level, which confirms our results of the importance of entropy for classifying the signals. We also selected a number of variance features.

Some studies have focused on analyzing ECG signals alone to detect stress. Medina perform clustering and dimensionality reduction on raw signals to determine whether the learned clusters corresponded to stress levels [18]. This author reports good results, which our study corroborates. Her results are not directly comparable to ours since she is not performing supervised learning but only unsupervised learning. Moreover, her data set is different from ours. Another study, by Sun et al. focused on detecting mental stress based on ECG signals [19]. Also using a different data set, therefore not directly comparable to our work, the authors report best classification accuracy results of 92.4% using decision trees. This study therefore confirms the capability of tree models to best discriminate between the features during the classification task. Differences with our study include using a different data set, training and test sets at 50% split, and using galvanic skin response in addition to ECG, which they report as increasing the classification accuracy rate. Other studies on using sensors for stress detection are summarized in a review paper [20].

6 Conclusion

Using machine learning algorithms from the ECG signals alone, we could achieve up to 100% accuracy and 1 AUC, with Multilayer Perceptron, depending on the experimental setting, in detecting three classes of stress: low, medium and high. Thus the accuracy of detecting multiple stress levels based on individual variations in ECG extracted features is higher than that of previously published results detecting stress based on fiducial points alone. These results were obtained by adding multiscale entropy measurements in addition to the fiducial measurements performed in previous studies on the same data set. Future work will include adding the T-wave related features in our analysis since ECG studies have shown that QT is an important biomarker of cardiac abnormality [21] and adding other signals. Clearly the results presented here are limited by the small size of the data set (67 samples) so that studies on larger data sets need to be conducted. We also plan to work with a physician for future directions of this work and to explore additional classification and clustering algorithms, for example hierarchical methods.

Nevertheless, the results of the present study lead to the exciting possibility of monitoring and diagnosing individual stress levels and alert the users accordingly so that accidents committed due to high or prolonged stress can be prevented. The personalized signal classification analysis presented here can be extended to other situations in which people face stress thereby addressing fatigue in workers in a factory, failure in functioning of the elderly people, players in a field, soldiers in a war field, etc.

References

1. Rubin, J., Eldardiry, H., Abreu, R., Ahern, S., Du, H., Pattekar, A., Bobrow, D.G.: Towards a mobile and wearable system for predicting panic attacks. In: Proceedings of the 2015 ACM International Joint Conference on Pervasive and Ubiquitous Computing, pp. 529–533. ACM, September 2015
2. Cruz, L., Rubin, J., Abreu, R., Ahern, S., Eldardiry, H., Bobrow, D.G.: A wearable and mobile intervention delivery system for individuals with panic disorder. In: Proceedings of the 14th International Conference on Mobile and Ubiquitous Multimedia, pp. 175–182. ACM, November 2015
3. Healey, J., Picard, R.W.: Detecting stress during real-world driving tasks using physiological sensors. IEEE Trans. Intell. Transp. Syst. 6(2), 156–166 (2005)
4. Rodrigues, J.G., Kaiseler, M., Aguiar, A., Silva Cunha, J.P., Barros, J.: A mobile sensing approach to stress detection and memory activation for public bus drivers. IEEE Trans. Intell. Transp. Syst. 16(6), 3294–3303 (2015)
5. Keshan, N., Parimi, P., Bichindaritz, I.: Machine learning for stress detection from ECG signals in automobile drivers. In: Proceedings of International IEEE Conference on Big Data, pp. 2449–2457. IEEE (2015)
6. Price, D.: How to read and electrocardiogram (ECG). Part 1: basic principles of the ECG. The normal ECG. South. Sudan Med. J. 3(2), 26–28 (2010)
7. Pan, J., Tompkins, W.J.: A real-time QRS detection algorithm. IEEE Trans. Biomed. Eng. 3, 230–236 (1985)

8. Goldberger, A.I., Amara, L.A.N., Glass, L., Hausdorff, J.M., Ivanov, P.C., Mark, R.G., Mietus, J.E., Moody, G.B., Peng, C.K., Stanley, H.E.: PhysioBank, PhysioToolkit, and PhysioNet: Components of a New Research Resource for Complex Physiologic Signals. Circulation **101**(23), e215–e220 (2010). doi:10.1161/01.CIR.101.23.e215. Circulation Electronic Pages: http://circ.ahajournals.org/cgi/content/full/101/23/e215. 13 June 2000. PMID: 10851218
9. Akbas, A.: Evaluation of the physiological data indicating the dynamic stress level of drivers. Sci. Res. Essays **6**(2), 430–439 (2011)
10. Hall, M., Frank, E., Holmes, G., Pfahringer, G., Reutemann, P., Witten, I.H.: The WEKA data mining software: an update. SIGKDD Explor. **11**(1), 10–18 (2009)
11. Costa, M., Goldberger, A.L., Peng, C.-K.: Multiscale entropy analysis of biological signals. Phys. Rev. **71**, 021906-1–021906-18 (2005)
12. Deng, Y., Wu, Z., Chu, C.-H., et al.: Evaluating feature selection for stress identification. In: IEEE Proceedings of the 13th International Conference on Information Reuse and Integration (IRI), pp. 584–591. IEEE (2012)
13. Deng, Y., Wu, Z., Chu, C.-H., et al.: An investigation of decision analytic methodologies for stress identification. Int. J. Smart Sens. Intell. Syst. (2012). ISSN: 1178-5608
14. Deng, Y., Wu, Z., Chu, C.-H., et al.: Sensor feature selection and combination for stress identification using combinatorial fusion. Int. J. Adv. Rob. Syst. **10**, 306–313 (2013)
15. Singh, M., Queyam, A.B.: A novel method of stress detection using physiological measurements of automobile drivers. Int. J. Electron. Eng. **5**(2), 13–20 (2013)
16. Singh, M., Queyam, A.B.: Correlation between physiological parameters of automobile drivers and traffic condition. Int. J. Electron. Eng. **5**(2), 6–12 (2013)
17. Avci, C., Akbas, A., Yüksel, Y.: Evaluation of Statistical Metrics by Using Physiological Data to Identify the Stress Level of Drivers
18. Medina, L.: Identification of stress states from ECG signals using unsupervised learning methods. Dissertation, Master's thesis, Universidade Técnica de Lisboa, Instituto Superior Técnico (2009)
19. Sun, F.T., Kuo, C., Cheng, H.T., Buthpitiya, S., Collins, P., Griss, M.: Activity-aware mental stress detection using physiological sensors. In: Mobile Computing, Applications, and Services, pp. 211–230. Springer, Heidelberg (2010)
20. Begum, S.: Intelligent driver monitoring systems based on physiological sensor signals: a review. In: Proceedings of the 16th International IEEE Conference on Intelligent Transportation Systems-(ITSC), pp. 282–289. IEEE (2013)
21. Zhou, Y., Sedransk, N.: Functional data analytic approach of modeling ECG T-wave shape to measure cardiovascular behavior. Ann. Appl. Stat. **3**, 1382–1402 (2009)

Smart Medical and Healthcare System 2017 Workshop

Augmenting Guideline Knowledge with Non-compliant Clinical Decisions: Experience-Based Decision Support

Naiara Muro[1,2,5(✉)], Nekane Larburu[1,2], Jacques Bouaud[4,5], Jon Belloso[3],
Gerardo Cajaraville[3], Ander Urruticoechea[3], and Brigitte Séroussi[5,6]

[1] eHeatlh and Biomedical Applications, Vicomtech-IK4, Donostia-San Sebastián, Spain
{nmuro,nlarburu}@vicomtech.org
[2] Biodonostia, Donostia-San Sebastián, Spain
[3] Onkologikoa Foundation, Donostia-San Sebastián, Spain
{jbelloso,gcajaraville,anderu}@onkologikoa.org
[4] AP-HP, DRCI, Paris, France
jacques.bouaud@aphp.fr
[5] Sorbonne Universités, UPMC, Univ Paris 06, INSERM, Université Paris 13,
Sorbonne Paris Cité, UMR S 1142, LIMICS, Paris, France
brigitte.seroussi@aphp.fr
[6] AP-HP, Hôpital Tenon, DSP, Paris, France

Abstract. Guideline-based clinical decision support systems (CDSSs) are expected to improve the quality of care by providing best evidence-based recommendations. However, because clinical practice guidelines (CPGs) may be incomplete and often lag behind the publication time of very last scientific results, CDSSs may not provide up-to-date treatments. It happens that clinical decisions made for specific patients do not comply with CDSS recommendations, whereas they comply with the state of the art. They may also be non-compliant because they rely on some implicit knowledge not covered by CPGs. We propose to capitalize the clinical know-how built from such non-compliant decisions and allow physicians to use it in future similar cases by the development of a decisional event structure that allows the modelling, storage, processing, and reuse of all the information related to a decision-making process. This structure allows the analysis of non-compliant decisions, which generates new experience-based rules. These new rules augment the knowledge embedded in CPGs supporting clinician decision for specific patients poorly covered by CPGs. This work is applied to the management of breast cancer within the EU Horizon 2020 project DESIREE.

Keywords: Experience-based clinical decision support system · Data mining techniques · Clinical guidelines evolution · Breast cancer · DESIREE

1 Introduction

Clinical practice guidelines (CPGs) are proposed as a source of information and treatment recommendations that rely on the rigorous evaluation of scientific publications to provide best health care practices [1]. However, CPGs have some weaknesses. The identification and synthesis of the evidence (e.g. deciding what type of evidence and

© Springer International Publishing AG 2018
Y.-W. Chen et al. (eds.), *Innovation in Medicine and Healthcare 2017*, Smart Innovation,
Systems and Technologies 71, DOI 10.1007/978-3-319-59397-5_23

outcomes should be included in guidelines), the determination of which values should be representative to be integrated in the guideline definition and how to update and implement these guidelines are some of them [2].

Most current clinical decision support systems (CDSSs) facilitate the implementation of CPGs [3], but they still do not overcome the weak points reflected above. For example, current CDSSs do not model implicit clinical knowledge not reflected in CPGs. Consequently, when clinical professionals perform the reasoning process that uses this implicit knowledge, and do not follow CPGs recommendations, i.e. when they make *non-compliant* decisions, the context and the reasoning process in which the implicit knowledge has been used are lost [4]. Over a 9-year period and more than 1000 breast cancer cases, Lin et al. [5] showed that actual chemotherapy decisions deviated from international guidelines in approximately 50% of the cases. This shows that CDSSs may end up useless for clinicians, since they use such systems to support them specially in the decision for special cases not addressed in CPGs.

Therefore, the main objective of our work is to store and process all the relevant information involved in the decision-making process of non-compliant decisions, to enrich the current CPG-based knowledge base formalized in the CDSS. The paper is organized as follows: Sect. 2 presents the state of the art about the main technologies used as basis of our work; Sect. 2.3 presents a new decision centered structure that will allow the exploitation of the information for each decisional event; Sect. 3 presents the methodology for generating new knowledge and a use case to illustrate it. Finally, Sect. 4 concludes the paper and proposes some future work.

2 Background Concepts

2.1 Clinical Practice Guidelines (CPGs)[1]

CPGs are defined as explicit and structured statements that model the current Evidence-Based Medicine (EBM) and the clinical judgment for best patient care at the decision making level [6, 7]. Good quality CPGs must present some characteristics including validity, reproducibility, reliability, representative development, clinical applicability, clinical flexibility and clarity [8]. Implementing CPGs has several benefits among which supporting clinicians in their decision making process, providing educational help for practitioners, improving quality assurance and assessment of the recommended treatment, and avoiding negligent medical practice [9].

Nevertheless, there are some barriers to the implementation and dissemination of CPGs that must be overcome to guarantee they are followed up in clinical practices. One of the main problems is the maintenance and update of CPGs, since because CPGs are not usually expressed in flexible and evolutive platforms, it often happens that CPG contents lag behind actual knowledge [10]. Furthermore, CPGs do not cover all possible clinical cases and recommendations for the specific patients that do not completely fit guideline contents, mainly due to these CPG knowledge gaps [11].

---

[1] In this paper, we refer to CPGs when talking about CPGs and local (validated) protocols.

2.2   Clinical Decision Support Systems (CDSSs)

In the last decade, CDSSs have proven to be potential tools to promote the implementation of CPGs [12–15] and give assistance to the clinicians in a decision-making process. They are often designed to help the implementation, integration, and application of CPGs, i.e. guideline-based CDSSs support clinicians in making CPG-compliant decisions [16]. Studies have reported that CDSSs do improve care quality and decrease medical errors [17]. Although guideline-based CDSS have a positive impact on the quality of medical practice they are quite constraining, as they depend on the *a priori* defined domain knowledge.

2.3   Techniques for Knowledge Discovery

Large biomedical databases contain unexploited knowledge that can give relevant information in the decision-making process.

Data mining techniques aim to discover this knowledge using classification, clustering and association algorithms [18]. In breast cancer domain for example data mining techniques are used mainly to predict the best result from a treatment for a patient [19] or to perform its survivability [20].

On the other hand, machine learning techniques, such as Case-Based reasoning (CBR) provides a recommendation for a new patient based on the decision previously made for similar patients [21]. Four steps are followed to get the recommendation: (i) case retrieval within the knowledge base built from previously solved cases, (ii) reuse of the most similar case(s), (iii) solution testing to see how the prior decision(s) fit(s) to the new case, and (iv) record of the newly acquired knowledge [22].

Current studies describe CPG implementation though different applications, such as rule based CDSS [2] or CBR [23, 24], along with data mining techniques to cover clinical "grey areas" that CPGs are not able to manage or for which their definition is relatively fuzzy [25].

3   Methods for Experience Modelling

Considering all the above mentioned constrains of current CDSS, we propose a new paradigm of decision support named "experience-based" as a hybrid CDSS following the principle of augmenting CPGs knowledge from data mining techniques and the study of CPG non-compliant cases.

This section describes the method we proposed to augment the current guideline-based CDSSs with experience, which results in an experience-based CDSS. For that, we first describe the decisional events structure that allows us to retrieve, model, and exploit all the information related to the decision-making process (Sect. 3.1). Thereafter we present the method to enrich CPGs by adding experience-based rules based on the decisional events information (Sect. 3.2).

3.1 Decisional Event Structure

A decisional event structure has been proposed in [26] to model all the information regarding the decision-making process. This decisional event structure is defined by a set of components:

1. $P = \{P_i\}$: Set of patient clinical parameters
2. $R = \{R_j\}$: Set of clinical statements expressed in a computer-interpretable way (IF-THEN rules). These clinical statements represent the knowledge coming from different sources (e.g. CPGs, local guidelines, experience-based rules generated by the system) and are itemized in the following components:
   (a) $A = \{A_m\}$: Set of the antecedents that compose the *conditional part* of rules, i.e. the IF-part. These antecedents evaluate patient clinical parameters with *a priori* defined conditions by CPGs with relational mathematic operators.
   (b) W: A recommendation coming from the accomplishment of the conditions defined in the antecedents, which is the *consequence part* of the rule, i.e. the THEN-part. In some cases, the provided recommendation could be an aggrupation of various treatments (i.e. a set of recommendations), expressed as
   $$W = \{S_1, S_2, \dots S_l\} \text{ with } l > 1 \text{ where } S \text{ is an atomic treatment.}$$
3. FD: Final decision taken by clinicians which could be compliant with the recommendation provided by the guideline W or not.
4. E: Actual treatment administrated at time t_1 after the decision is made, which could be compliant with FD or not.
5. $C = \{C_k\}$: Set of criteria followed by clinicians to reach an agreement about a final decision. These criteria will be sorted in different groups that will have a closed list of Boolean possible values J_n. So, we can define a single criterion as a set of justifications $C_1 = \{J_1, J_2, \dots J_n\}$ with $n \geq 1$. For example, *Tumor Size* could be a criterion of non-compliance which justification is the difficult follow-up, which could be either *true* or *false* (i.e. Boolean value).
6. $O(t)$: Set of outcomes of a studied patient after a time t to be able to assess the success or failure of the given treatment.

Ideally, clinicians' decisions are compliant with CPGs, thus choosing one of the recommendation provided by the guideline-based CDSS as their final decision FD. But in certain cases, when clinicians do not comply with CPGs (e.g. BU considers the patient preferences in their decision), FD is different from CPG-based recommendation(s) for that patient. In both cases, the administrated treatment E is expected to be equal to the final decision FD, but due to deviations in the treatment plan, E could differ from FD.

The modelling of all the contextual information of a decision-making process into a decisional event structure makes possible to understand, process, and reuse the implicit clinical knowledge.

3.2 Experience-Based Rules

The data modeled within the decisional event structure is used to identify relevant information in the decision-making process and to retrieve implicit clinical knowledge [6]. In cases where clinicians do not follow CPGs-based recommendation(s), thus being non-compliant with the CPGs, there is an implicit clinical knowledge that we seek to exploit to enrich the knowledge base of the CDSS.

Below we present the method to analyse a decisional event and build the experience-based rules from non-compliant decisions:

The starting point of the method is to retrieve the set of CPG rules that were executed in the decisional event $RS = \{R_1, R_2, \ldots R_j\}$ $with\, j > 0$. The antecedents (i.e. IF-part) of these rules are defined as $CR = \{A_1, A_2, \ldots A_k\}$ $with\, k > 0$. The evaluated patient parameters accomplished all of them.

Thereafter, we identify and retrieve the rule set $RS' = \{R'_1, R'_2, \ldots R'_u\}$ $with\, u > 0$ whose recommendation W match with the final decision FD made by clinicians. The antecedents of this secondary rule set RS' are defined as $CR' = \{A'_1, A'_2, \ldots A'_m\}$ $with\, m > 0$ and the evaluated patient parameters did not accomplish at least one from each rule R'.

From both sets of antecedents, CR and CR', we look for *'conflictive antecedents'*, i.e. incompatible antecedents (e.g. *Tumor Size >20* ϵ CR and *Tumor Size ≤20* ϵ CR'), or antecedents that are complementary, i.e. *'complementary antecedents'* (See example in the Use Case explained below).

We keep the complementary antecedents in the experience-based rule generated from the non-compliant decision.

- In some cases, one or more antecedent could be defined in the non-compliant antecedent set CR' but not in the compliant one CR, i.e. antecedents defined in the relative complement of CR formally noted as: $CR' \backslash CR$. For this scenario, this new antecedent will be included in the new rule with the patient's clinical parameter as constraint value.

When the identified CR and CR' sets contain conflictive antecedents, the following steps must be adopted, depending on the reasons of the conflict:

- If the antecedent is defined in both CR and CR' (i.e. when it is defined in $CR \cap CR'$) but with different value constraints, in the new experience-based rule, this antecedent will be defined with the patient's clinical parameter value as constraint (e.g. the tumor size in CR is characterized by *Tumor Size >20* whereas for CR' is measured as *Tumor Size ≤20 CR'*. In the experience based rule it will take the patient value: *Tumor Size = P_i*). In this case, we will be adjusting the value of a constraint.

Lastly, the set of criteria C_k (e.g. clinical preferences, patient preferences) defined by clinicians in the decision-making process composed by one or more Boolean justifications J_n give us hints about new relevant clinical parameters to include (e.g. because they were not defined in the CPGs) or study.

To sum up, when a new clinical parameter has to be added in the generation of the experience based rule from one of the studied rule sets CR or CR' (i.e. complementary antecedents) it must always be equal to the patient's value.

To illustrate the applicability of this method, a use case is presented next.

3.3   Use Case: Breast Cancer

We present a simplified use case based on the local protocol from Onkologikoa Foundation, where we apply the previously presented method. We consider two patients, Patient 1 and Patient 2 suffering from an Invasive Ductal Carcinoma (IDC).

The highlighted parameters (in grey) are those considered by Onkologikoa's protocols. In Fig. 1 we illustrate one of the rules from which the antecedents of its conditional statement are met for patient 1.

```
R {
        IF {
            A1   (P14 = "Ductal invasive carcinoma") AND
            A2   (P17 = Positive) OR
            A3   (P18 = Positive) AND
            A4   (P19 = Negative) AND
            A5   (P20 < 20) AND
            A6   (P6 > 20)
        }                                                    CR

        THEN {
                (Recommendation = "Neo-Adjuvant Hormonotherapy")
        }                                                    W
```

Fig. 1.   Local protocol derived rule for non-metastatic breast cancer with infiltrating tumor

In Fig. 1 we identify the conditional statement (in blue) named CR. This conditional statement is composed by a set of antecedents (i.e. $CR = A_1 \cup A_2 \cup A_3 \cup A_4 \cup A_5 \cup A_6$, highlighted in green). The consequence statement (in orange) provides protocol-based recommendation $W =$ "Neo-Adjuvant Hormonotherapy".

The BU decided not to comply with the provided recommendation and decided $FD =$ "surgery". The reason behind this final decision was the criteria $C =$ "Tumor Size" with the justification $J_1 =$ "Follow-up difficulty".

In Fig. 2 we summarize the data related to this decisional event. The criteria C and justification J that explained the decision of the different treatment and the non-compliant FD are highlighted because their source was not protocol-based, but relied on clinicians' know-how.

```
Decisional Event (PATIENT 1)

    •  Clinical Parameters: {P_i} with 1 ≤ i ≥ 20
       (i.e. P1: Age = 65, P2: Sex = Woman, ...)
    •  Clinical Condition: R
          o  Set of Antecedents: {A_m} with 1 ≤ m ≥ 6
             (i.e. A1: P14 = "Ductal invasive carcinoma", A2: P17 = Positive, ...)
          o  Recommendation: W = "Neo – Adjuvant Hormonotherapy"
    •  Clinical criterion: C = "Tumor Size"
          o  Justification: J1 = "Follow – up difficulty"
    •  Final decisión: FD = "Surgery"
```

Fig. 2. Summary of the data that composes the decisional event for Patient 1

Once the decision-making process is completed, and since the decision was not compliant, data is processed to retrieve the implicit knowledge used and consequently augment the knowledge base. The set of rules which recommendation W matches with FD = "surgery" is retrieved. Figure 3 shows a protocol rule that does not match Patient 1 clinical parameter "size" (highlighted in red in the figure), but provides the desired recommendation "surgery".

```
IF {
     (P14 = "Ductal invasive carcinoma") AND
     (P17 = Positive) OR
     (P18 = Positive) AND
     (P19 = Negative) AND
     (P6 < 20) AND
     (P11 = 0)
}
THEN {
     (Recommendation = "Surgery")
}
```

Fig. 3. Example of a protocol rule that provides the desired recommendation despite it does not apply to Patient 1

The new experience-based rule (Fig. 4) will contain **(i)** the antecedents checked by both rules, i.e. *'complementary antecedents'* $\in CR \cap CR'(same\ P_i\ equal\ value)$ (in black), **(ii)** the adjustment of the parameter that was not compliant in one of them characterized by the most restrictive value, i.e. *'conflictive antecedents'* $\in CR \cap CR'\left(same\ P_i\ different\ value\right)$ (in blue), **(iii)** the inclusion of a clinical parameter that was only measured in one of the rules, i.e. inclusion of $P_i \in CR \backslash CR' \cup CR' \backslash CR$ (in green) and **(iv)** the inclusion of clinical criteria that justifies the non-compliancy from the BU (in orange):

```
IF {
     (P₁₄ = "Ductal invasive carcinoma") AND
     (P₁₇ = Positive) OR
     (P₁₈ = Positive) AND
     (P₁₉ = Negative) AND
     (P₂₀ <20) AND
     (P₆ = 21) AND
     (P₁₁ = 0)
}
THEN {
     (Recommendation = "Surgery"
      Justification= "Tumor Size: Follow-up difficulty")
}
```

Fig. 4. The experience-base rule generated from the non-compliant decision for Patient 1

Once the experience-based rule is generated, it will be stored in the knowledge base and could be fired for any patient whose clinical parameters checked the conditional statement. To illustrate such case, we present Patient 2 (Table 1). Notice that Patient 2 has parameters similar as those defined for Patient 1 (Table 1), but some additional information concerning clinical parameters (P_{21}: *Tumor size = Follow-up difficulty*).

Table 1. Set of clinical parameters and values defining Patient 1 and Patient 2

		PATIENT 1	PATIENT 2
P_1	Age	65	73
P_2	Sex	Woman	Woman
P_3	Number of Pregnancies	0	4
P_4	Number of Lesions	2	2
P_5	Location	Right, Lower outer quadrant	Right, Upper outer quadrant
P_6	Size (mm)	21	22
P_7	BIRADS	5	-
P_8	Ulceration	NO	NO
P_9	Skin metastasis	NO	NO
P_{10}	cT (size)	T2	T2
P_{11}	cN (number)	0	0
P_{12}	cM (metastasis)	0	0
P_{13}	Stage	2a	2a
P_{14}	Histological type	Invasive	Invasive
P_{15}	Grade	GII	GI
P_{16}	Carcinoma in situ type	Ductal carcinoma	Ductal carcinoma
P_{17}	Estrogen receptor	Positive	Positive
P_{18}	Progesterone receptor	Positive	Positive
P_{19}	HER-2 receptor	Negative	Negative
P_{20}	Ki67 (%)	14	16
P_{21}	Clinical criterion: Tumor Size	-	Follow-up difficulty

For Patient 2, protocol- and experience-based rules are executed and provide the two recommendations displayed in Fig. 5.

```
CPG Recommendation: Neo-Adjuvant Hormonotherapy
Experience Recommendation: Surgery
(Justification = Tumor Size: Follow-up difficulty)
```

Fig. 5. Recommendations generated by both protocol- and experience based rule sets for Patient 2

4   Conclusions and Future Work

This work presents a methodology to augment the knowledge of CPGs with clinician experience. First, a decisional event structure is described. This structure formalizes all the decision-related parameters in a computer-interpretable way, allowing its interpretation and reuse. The decisional event structure includes data that plays an important role in the decision-making process (e.g. patients preferences, clinician preferences...), but is not explicitly considered in current CPGs and often explains the reason of non-compliance with CPGs. Hence, this decisional event structure can be a source of knowledge discovery and a starting point for the study of CPG update to cover uncovered specific clinical cases (e.g. onco-geriatric cases).

Second, based on the decisional event structure, we presented the exploitation of the events related to cases where clinicians do not comply with CPGs. For that, we analyzed and processed the implicit clinical knowledge, often omitted in current clinical daily practices, that affects the decision-making process. This process allows the creation of new experience-based rules, which are part of CPGs evolvement.

Nevertheless, the generated experience-based rules, generated from non-compliant cases, must be validated by clinicians to include them in the CPGs-based rule set. This way, we avoid polluting the CPG knowledge base when adding new rules, without clinical supervision and acceptance.

As future work, we will build a quality assessment algorithm that will provide information about the success or failure of the treatments recommended by experience-based rules based on different parameters defined by the outcomes of the patient, such as quality of life or life expectancy.

Acknowledgements. This project has received funding from the European Union's Horizon 2020 research and innovation program under grant agreement No 690238.

References

1. Grimshaw, J.M., Russell, I.T.: Achieving health gain through clinical guidelines II: ensuring guidelines change medical practice. Qual. Health Care 3(1), 45–52 (1994)
2. Woolf, S., Schünemann, H.J., Eccles, M.P., Grimshaw, J.M., Shekelle, P.: Developing clinical practice guidelines: types of evidence and outcomes; values and economics, synthesis, grading, and presentation and deriving recommendations. Implement. Sci. **7**, 61 (2012)
3. Kawamoto, K., Houlihan, C.A., Balas, E.A., Lobach, D.F.: Improving clinical practice using clinical decision support systems: a systematic review of trials to identify features critical to success. BMJ **330**(7494), 765 (2005)

4. Galanter, W.L., Didomenico, R.J., Polikaitis, A.: A trial of automated decision support alerts for contraindicated medications using computerized physician order entry. JAMIA **12**(3), 269–274 (2005)
5. Lin, F.P.Y., Pokorny, A., Teng, C., Dear, R., Epstein, R.J.: Computational prediction of multidisciplinary team decision-making for adjuvant breast cancer drug therapies: a machine learning approach. BMC Cancer **16**(1), 929 (2016)
6. Lobach, D.F., Hammond, W.E.: Computerized decision support based on a clinical practice guideline improves compliance with care standards. Am. J. Med. **102**(1), 89–98 (1997)
7. Sackett, D.L., Rosenberg, W.M.C., Gray, J.A.M., Haynes, R.B., Richardson, W.S.: Evidence based medicine: what it is and what it isn't. BMJ **312**(7023), 71–72 (1996)
8. Thomas, L.: Clinical practice guidelines. Evid. Based Nurs. **2**(2), 38–39 (1999)
9. Silberstein, S.: Clinical practice guidelines. Cephalalgia **25**(10), 765–766 (2005)
10. Wang, D., et al.: Representation primitives, process models and patient data in computer-interpretable clinical practice guidelines. Int. J. Med. Inf. **68**(1), 59–70 (2002)
11. Bates, D.W., et al.: Ten commandments for effective clinical decision support: making the practice of evidence-based medicine a reality. J. Am. Med. Inform. Assoc. **10**(6), 523–530 (2003)
12. Sim, I., et al.: Clinical decision support systems for the practice of evidence-based medicine. JAMIA **8**(6), 527–534 (2001)
13. IOS Press Ebooks - Computer-based Medical Guidelines and Protocols: A Primer and Current Trends. Accessed 09 Mar 2017
14. Foundations of biomedical knowledge representation - Google Search. Accessed 09 Mar 2017
15. Peleg, M.: Computer-interpretable clinical guidelines: a methodological review. J. Biomed. Inform. **46**(4), 744–763 (2013)
16. Berner, E.S.: Clinical Decision Support Systems. Springer, New York (2007)
17. Berner, E.S., Lande, T.J.L.: Overview of clinical decision support systems. In: Berner, E.S. (ed.) Clinical Decision Support Systems, pp. 1–17. Springer International Publishing (2016)
18. Yoo, I., et al.: Data mining in healthcare and biomedicine: a survey of the literature. J. Med. Syst. **36**(4), 2431–2448 (2012)
19. Xiong, X., Kim, Y., Baek, Y., Rhee, D.W., Kim, S.-H.: Analysis of breast cancer using data mining statistical techniques. In: Proceedings of the 6th SNPD/ACIS, pp. 82–87 (2005)
20. Sarvestani, A.S., Safavi, A.A., Parandeh, N.M., Salehi, M.: Predicting breast cancer survivability using data mining techniques. In: 2010 Proceedings of the 2nd International Conference on Software Technology and Engineering, vol. 2, pp. V2-227–V2-231 (2010)
21. Frize, M., Walker, R.: Clinical decision-support systems for intensive care units using case-based reasoning. Med. Eng. Phys. **22**(9), 671–677 (2000)
22. Aamodt, A., Plaza, E.: Case-based reasoning: foundational issues, methodological variations, and system approaches. AI Commun. **7**(1), 39–59 (1994)
23. Montani, S.: Case-based reasoning for managing noncompliance with clinical guidelines. Comput. Intell. **25**(3), 196–213 (2009)
24. D'Aquin, M., Lieber, J., Napoli, A.: Adaptation knowledge acquisition: a case study for case-based decision support in oncology. Comput. Intell. **22**(3–4), 161–176 (2006)
25. Toussi, M., Lamy, J.-B., Le Toumelin, P., Venot, A.: Using data mining techniques to explore physicians' therapeutic decisions when clinical guidelines do not provide recommendations: methods and example for type 2 diabetes. BMC Med. Inform. Decis. Mak. **9**, 28 (2009)
26. Larburu, N., Muro, N., Macía, I., Sánchez, E., Wang, H., Winder, J., Bouaud, J., Séroussi, B.: Augmenting guideline-based CDSS with experts' knowledge. In: HealthInf (2017)

Upgrading Legacy EHR Systems to Smart EHR Systems

Ane Murua[1,2(✉)], Eduardo Carrasco[1,2], Agustin Agirre[3],
Jose Maria Susperregi[3], and Jesús Gómez[3]

[1] Vicomtech-IK4, Donostia, San Sebastián, Spain
{amurua,ecarrasco}@vicomtech.org
[2] Biodonostia Health Research Institute, San Sebastian, Spain
[3] La Asunción Clinic, Tolosa, Spain
{aagirre,jsusperregi,jesusgm}@clinicadelaasuncion.com

Abstract. Electronic Health Record (EHR) systems are a key element of the clinical practice in most hospitals and healthcare organizations. Although traditionally its role has been focused mainly as a patient health data storage and communication tool, thanks to the recent technical advancements, a wide range of new promising possibilities are arising, including, user empowerment, new medical knowledge discovery, clinical decision support systems and clinical tasks automation. This paper discusses a set of different features that can be added to a legacy EHR system to upgrade it into a Smart EHR system, such as (i) health related data curation, (ii) rule-based data processing, (iii) business process management and (iv) intelligent agents.

Keywords: Electronic Health Records · Personal health records · Knowledge engineering · Rule-based systems · Business process management · Natural language processing · Intelligent agents

1 Introduction

The management of large amount of patient information in medical practice has made the medical record the cornerstone of clinical communication and documentation [1]. This patient information was stored in the form of paper based medical record entirely until early 1960s when the idea of electronic medical record was introduced [2] and progressively extended since then.

The implementation and adoption of EHR systems throughout the world differ in developing and developed countries [3]. The developing countries are starting to implement EHR systems as supporters of paper-based health records [4], while many developed countries have nationwide policies to foster EHR adoption. In several countries (e.g. New Zealand, Sweden, Norway, Netherlands, United Kingdom, Australia or the United States), the percentage of primary care physicians using electronic medical records is almost 100% [5]. In Spain, this percentage is higher than 90%, but the situation is mixed: while older hospitals are facing implantation issues, new hospitals have complete EHR systems [6], and some hospitals also develop and market their own EHR system, such as Hygehos EHR [7].

Y.-W. Chen et al. (eds.), *Innovation in Medicine and Healthcare 2017*, Smart Innovation, Systems and Technologies 71, DOI 10.1007/978-3-319-59397-5_24

Literature shows that EHR systems provide relevant benefits for clinical outcomes (e.g., improved quality, reduced medical errors), organizational outcomes (e.g., financial and operational benefits), and societal outcomes (e.g., improved ability to conduct research, improved population health, reduced costs). Similarly, several important drawbacks can be identified as well such as the high upfront acquisition and maintenance costs, disruptions to clinical workflows, losses in productivity in the learning stages [8]. Nevertheless, it is agreed that significant benefits are brought to patients and society when EHR systems are used in an appropriate way.

2 Next-Generation EHR Systems

Due to the relevance of EHR systems in our society, intense research and development has been conducted and new definitions are starting to emerge. In this sense, according to [9] an Electronic Health Record (EHR) system includes: (1) longitudinal collection of electronic health information for and about persons, where health information is defined as information pertaining to the health of an individual or health care provided to an individual; (2) immediate electronic access to person- and population-level information by authorized, and only authorized, users; (3) provision of knowledge and decision-support that enhance the quality, safety, and efficiency of patient care; and (4) support of efficient processes for health care delivery.

New information and communication technologies have the potential to fully meet and extend the aforementioned goals, but, in order to do so, several relevant technical challenges have to be addressed.

First, the data contained in the EHRs has to be further structured and codified using standardized terminologies. Experts agree on that 80% of the data in the healthcare sector is unstructured, and hence no further exploited [10].

Next, patient health data is expected to continue growing exponentially in the coming years and a great part of all these data will be physically scattered beyond the limits of the healthcare organizations. The management of all this information and the access and privacy issues involved raises new challenges for the EHR systems [11].

Third, available healthcare data should be transformed into reusable knowledge, and this knowledge will be confronted with stablished clinical guidelines in order to discover the best practices that lead to the best decision support given to the medical practitioners [12].

Finally, medical workflows have to be automated in the EHR systems as much as possible in order to increase the efficiency of the health care delivery [13].

Relevant EHR systems have been developed in the last years that are aligned with these challenges such as Kaiser Permanente [14] or openEHR [15], but still progress has to be carried out in order to meet them at their full extent.

3 Methods for Upgrading Legacy EHR Systems

In this section, several technologies that can be integrated into legacy EHR systems in order to upgrade them into Smart EHR systems are described.

3.1 Data Curation

The first milestone for upgrading a legacy EHR system is to improve the quality of health data contained in it, to foster its reuse, to add more value to the data and to add complementary sources of data.

Natural Language Processing Techniques

As stated above in this paper, 80% of the data in the healthcare sector is in unstruc-tured formats which include machine-written, handwritten information and audio dictations among others. In these formats, relevant health information is "locked" since they were intended only for human reading and interpretation.

Extracting key data elements from unstructured medical records into structured computable data elements is an essential step for EHR information reuse. Natural Language Processing techniques are a straightforward tool to help automating this task. Numerous researchers and academic organizations have been exploring over the last decade the potential of natural language processing for risk stratification, population health management, and decision support. A recent example of a machine learning NLP in the healthcare industry is IBM Watson, which has been focused in clinical decision support for precision medicine and cancer care [16, 17]

In order to facilitate the computer-assisted extraction and understanding of the most relevant terms of the EHR records, available terminologies such as SNOMED CT or ICD 10 are used. UMLS [18] is remarkable as well, since it integrates and relates most relevant biomedical vocabularies available so far.

Standardized Data Models

Data documented in structured patient records is required to be uniformly coded and documented in order to be reliable and interoperable for utilization in direct patient care and in other contexts such as secondary use purposes [19].

During the last decade, different EHR standards had been developed for EHR modelling. The most extended ones are (i) HL7, (ii) ISO EN13606 and (iii) openEHR.

Although each standard brings a differential feature compared to others, all agree on a dual model structure, consisting on a reference model (RM) and an Archetype Model (AM).

The RM supports information within a structure, based on well-established concepts independent from knowledge. It represents the characteristics of the general components and their organization. AM defines and models concepts of clinical knowledge following the structure and constraints imposed by the RM.

The combination of both models in a single frame provides of structure and semantic interpretation to the content stored in the EHR [20].

Integration of Distributed Patient Data

According to an American study carried out by the "California HealthCare Foundation" in 2013 [21], 7 out of 10 adult Americans regularly measure at least one of their health status related indicators, such as their weight, diet or activity level.

A gradual transformation is occurring, causing the individual -who used to be a passive "element", merely an information generator-, to become a subject capable of analyzing its own data and even able to start acting according to the insights gained from its self-tracking [22].

Patient's own health and lifestyle data is meaningful for health organizations and should be integrated with the EHR in the form of a Personal Health Record (PHR) [23]. DocToDoor platform [24] is an example of the integration of PHR and EHR.

Besides the data annotated by the patient himself, PHR could contain data from other sources, such as: data gathered from sensors or other wearable computing devices, data acquired through mainstream smartphone applications' APIs, activity in social networks or current communication channels.

3.2 Rule-Based Systems

In a rule-based system the key idea is to separate knowledge and represent it as facts and rules, that is, as conditional sentences relating statements of facts with one another [25]. A rule-engine (e.g. Drools, CLIPS, OpenRules, JESS) provides an alternative computational model which can take rules (declared as a group of "if-then" statements) and execute them over data.

When the conditions stated in a rule are met ("if"), the rule is evaluated ("then") and our facts are updated accordingly.

The information contained on the EHR and PHR can be processed in a rule-based system to provide, according to [26], patient-specific, situation-specific alerts, reminders, or other recommendations for direct action; and to organize and present information in a way that facilitates problem solving and decision making using appropriate visual analytics technics.

3.3 Business Process Management

Business Process Management (BPM) is a methodology which describes the whole life cycle of how to discover, formalize, execute, and monitor our business processes. Business process models are modelled using predefined notations such as BPMNv2 [27], defined by the OMG group. These models have a graphical diagram showing the exact sequence of the activities that are going to be executed and include activities performed by both people and computers.

The most common uses of the BPM methodology in the healthcare field are:

- Computer-interpretable clinical guideline modelling.
- Health center's own (clinical, management, supply-chain, …) workflow orchestration.
- Implementing structured multidisciplinary care plans that detail essential steps in the care of patients with a specific clinical problem (care pathways)

Business processes can be used as a standalone tool or combined with rule-based systems. The most common patterns of rule and process integration are: (1) Including in a process a specific type of task called a Business Rule Task where a Rule Engine is called with some data to get some results; (2) Using rules to start processes to deal with different scenarios difficult to perform by just chaining rules; and (3) Inserting our Process Instances as facts, among some other facts, in a Rule Engine.

3.4 Intelligent Agents

Intelligent Agents are typically described as autonomous artificial entities that sense the world on a continuous basis and act (proactively or reactively) on it in order to achieve specific tasks, such as event detection, maintenance of a domain knowledge model, learning from their observations to improve performance [28].

These agents run in Multi-Agent Systems MAS which are integrated with the information systems in many different sectors. The integration of MAS with the EHR could lead into more advanced scenarios, i.e. applying intelligent agents in the healthcare domain with a wide range of applications, such as:

- *Data-management systems.* The focus is on the efficient retrieval and processing of scattered medical data, for example combining patients' data in the EHR with other sources such as most recent evidences available for treatment.
- *Decision Support Systems (DSSs).* DSSs provide patient-specific recommendations based on previous healthcare processes or knowledge-based models for example for diagnosis or treatment selection.
- *Planning.* Systems centered on the coordination and scheduling of human and material resources, for example when executing a standardized clinical guideline.
- *Simulation.* Agents can be used to make rule-based simulations of the behavior of complex challenges, such as to evaluate the impact of particular treatment taking into account the evolution of a disease.
- *Monitoring and alarms.* The goal is to continuously monitor the current state of a patient and, considering the evolution and the general context, warn the patient (or a supervisor) about problematic future situations.

4 Conclusions

EHR systems bring significant benefits to patients and society as they contribute, amongst others, to share medical information, to reduce medical errors, to improve coordination of care and health care quality and to lower national health care costs.

This work describes a set of methods focused on upgrading legacy EHR systems. This is a common scenario for EHR system development organizations who want to benefit from state-of-the-art technologies to explore new horizons.

The methods addressed in this paper focus on improving the quality of the health data in the EHR, its exploitation and computer reuse, the extraction of medical knowledge from health data, the promotion of computer-aided decision support and the automatization of clinical processes in order to promote efficiency in the healthcare provision. In particular, the technical topics addressed are (i) data curation, (ii) medical rules management, (iii) business process modeling and (iv) intelligent agents. The integration of these technologies in the EHR systems will pave the way to a new generation of Smart EHR systems.

Smart EHR systems will bring new appealing scenarios to the healthcare organizations, such as (i) empowerment of both patients and health professionals, (ii) reuse of health information in order to discovery of new medical knowledge, (iii) promote the development of Clinical Decision Support Systems to enhance the quality of the health-care, (iv) increase the sustainability and security by means of the automatization of clinical tasks and (v) advance to a Real-world evidence based medicine.

Acknowledgements. The authors wish to thank "Centro para el Desarrollo Tecnológico Industrial" (CDTI) for partially funding current R&D activities by means the project SMARTHEALTH – "Large Population Health Monitoring and Clinical Assessment Platform" (IDI-20160260) inside the Indo-Spanish Joint Programme of Co-operation on Industrial Research and Development between GITA India & CDTI Spain.

References

1. John, S.L.: Electronic medical records. Prim. Psychiatry **13**(2), 20–23 (2006)
2. Blumenthal, D., Tavenner, M.: The "Meaningful Use" regulation for electronic health records. N. Engl. J. Med. **363**, 501–504 (2010)
3. Al-Aswad, A.M., Brownsell, S., Palmer, R., Nichol, J.P.: A review paper of the current status of electronic health records adoption worldwide: the gap between developed and developing Countries. J. Health Inf. Dev. Countries **7**(2), 153–164 (2013)
4. Oak, M.R.: A review on barriers to implementing health informatics in developing countries. J. Health Inf. Dev. Countries **1**(1), 19–22 (2007)
5. Mossialos, E., Wenzl, M., Osborn, R., Sarnak, D. (eds.) International Profiles of Health Care Systems, 2015, The Commonwealth Fund, January 2016 (2016)
6. Carnicero, J., Rojas, D.: La explotación de datos de salud: Retos, oportunidades y límites. In: Carnicero, J., Rojas, D. (eds.) La explotación de datos de salud: Retos, oportunidades y límites. Sociedad Española de Informática de la Salud, Pamplona (2016)
7. Hygehos S.L. http://www.hygehos.com/hygehos-2-0
8. Menachemi, N., Collum, T.H.: Benefits and drawbacks of electronic health record systems. Risk Manag. Healthc. Policy **4**, 47–55 (2011)
9. Institute of Medicine: Key Capabilities of an Electronic Health Record System: Letter Report. The National Academies Press, Washington, DC (2003)
10. Unstructured Data in Electronic Health Record (EHR) Systems: Challenges and Solutions. Datamark (2013)
11. Brown, N.: Healthcare Data Growth: An Exponential Problem. Nextech (2015)
12. Larburu, N., et al: Augmenting guideline-based CDSS with experts' knowledge. In: HealthInf 10th International Conference on Health Informatics (2017)
13. Gooch, P., Roudsari, A.: Computerization of workflows, guidelines, and care pathways: a review of implementation challenges for process-oriented health information systems. J Am. Med. Inf. Assoc. **18**(6), 738–748 (2011)
14. Liang, L.L., Berwick, D.M.: Connected for Health: Using Electronic Health Records to Transform Care Delivery. Kaiser Permanente, Jossey-Bass, San Francisco (2010)
15. Atalog, N., et al.: openEHR. A semantically-enabled, vendor-independent health computing platform. White paper (2017)
16. Nadkarni, P.M., et al.: Natural language processing: an introduction. J. Am. Med. Inform. Assoc. **18**(5), 544–551 (2011)
17. Doyle-Lindrud, S.: Watson will see you now: a supercomputer to help clinicians make informed treatment decisions. Clin. J. Oncol. Nurs. Tech Savvy **19**(1), 31 (2015)
18. Bethesda: UMLS® Reference Manual. National Library of Medicine (US) (2009)

19. Vuokko, R., Mäkelä-Bengs, P., Hyppönen, H., Lindqvist, M., Doupi, P.: Impacts of structuring the electronic health record: Results of a systematic literature review from the perspective of secondary use of patient data. Int. J. Med. Inf. **97**, 293–303 (2017)
20. Muro, N., Sanchez, E., Graña, M., Carrasco, E., Manzano, F., Susperregi, J.M., Agirre, A., Gómez, J.: Hygehos ontology for electronic health records. In: Chen, Y.W., Tanaka, S., Howlett, R., Jain, L. (eds.) Innovation in Medicine and Healthcare 2016. Smart Innovation, Systems and Technologies, vol. 60. Springer, Cham (2016)
21. Fox, S., Duggan, M.: Tracking for Health. Pew Research Center (2013)
22. Alvarez, R., Murua, A., Artetxe, A., Epelde, G., Beristain, A.: A platform for user empowerment through self ecological momentary assessment/intervention. In: MOBIHEALTH 2015 Proceedings of the 5th EAI International Conference on Wireless Mobile Communication and Healthcare, pp. 206–209 (2016)
23. Roehrs, A.: Personal health records: a systematic literature review. J. Med. Internet Res. **19**, 1 (2017)
24. DocToDoor features. http://doctodoor.com/features.html
25. Buchanan, B.G., Duda, R.O.: Principles of rule-based expert systems. Adv. Comput. **22**, 163–216 (1983)
26. Musen, M.A., Middleton, B., Greenes, R.A.: Clinical decision-support systems. In: Shortliffe, E.H., Cimino, J.J. (eds.) Biomedical Informatics. Springer, London (2014)
27. Object Management Group: Business Process Model and Notation. http://www.omg.org/spec/BPMN/
28. Isern, D., Moreno, A.: A systematic literature review of agents applied in healthcare. J. Med. Syst. **40**, 40–43 (2016)

Towards a Deconstructed PACS-as-a-Service System

Roberto Álvarez[1,2(✉)], Jon Haitz Legarreta[1,2], Luis Kabongo[1,2],
Gorka Epelde[1,2], and Iván Macía[1,2]

[1] eHealth and Biomedical Applications, Vicomtech-IK4,
Donostia-San Sebastián, Spain
{ralvarez, jhlegarreta,
lkabongo, gepelde, imacia}@vicomtech.org
[2] Biodonostia, Donostia-San Sebastián, Spain

Abstract. Traditional Picture Archiving and Communication Systems (PACS) were designed for vendor-specific environments, dedicated radiology workstations and scanner consoles. These kinds of systems are becoming obsolete due to two main reasons. Firstly, they don't satisfy the long-standing need in healthcare to put all the resources related to the patient into a single solution rather than a multitude of partial solutions. And secondly, communication, storage and security technologies have demonstrated that they are mature enough to support this demand in other fields. "Vendor Neutral Archives" are becoming the new trend in medical imaging storage and "deconstructed PACS" goes one step beyond proposing a totally decoupled implementation. Our work combines this implementation with the scalability and ubiquitous availability of cloud solutions and internet technologies to provide an architecture of a PACS-as-a-service system that handles a simple enterprise workflow orchestration of tele-radiology.

Keywords: Cloud-based technologies · Deconstructed PACS · DICOM · RSNA · Tele-radiology · VNA

1 Introduction

Traditional medical image archiving systems have been closed, vendor-dependent solutions, requiring non-negligible initial investment and additional maintenance costs. Such setups are commonly designated by the general term Picture Archiving and Communication Systems (PACS) [1], they typically comprise an infrastructure of data acquisition, storage and visualization systems integrated into a digital network, as well as the necessary software to provide such services.

According to [2], one of the biggest problems with traditional PACS arises when the hospital needs to migrate from one PACS vendor to another, typically motivated by the exponential growth and increased complexity of information. Since each solution has a different proprietary implementation of archiving strategies, migrations entail that the complete earlier data needs to be adapted to the format of the newly procured PACS, which is resource consuming. Furthermore, despite that these solutions are

© Springer International Publishing AG 2018
Y.-W. Chen et al. (eds.), *Innovation in Medicine and Healthcare 2017*, Smart Innovation, Systems and Technologies 71, DOI 10.1007/978-3-319-59397-5_25

designed to handle large amounts of data, storage demand is growing enormously, which implies a continuous and increasing number of upgrades in storage capabilities and configurations.

Statistics report that the average number of radiology scans is over 1 per year per inhabitant (different years reported), and trends show increasing figures [3]. Also, it was estimated that 1 billion diagnostic imaging procedures were performed in the United States in 2014, adding up 100 petabytes of volume data [4].

Medical image requirements are heterogeneous across medical specialties and clinical centers. In this context, structured and consistent digital data availability to physicians potentially avoids data loss, scan repetitions or delayed diagnosis. The use of modern web-based data storage, communication and image visualization technology, along with the integration of the Electronic Health Record (EHR) as key element, has the potential to transform and improve new solutions beyond PACS capabilities. Therefore, a pressing need for increasingly flexible and efficient solutions is arising.

This paper proposes an implementation that aims to avoid problems derived from the departmental focus of traditional approaches, which tends to create silos of information. The remainder of the document is structured as follows: Sect. 2 presents a summary of related research and development works, as well as existing commercial products, and other relevant aspects; Sect. 3 presents the specific context the work is aimed at; Sect. 4 provides an overview of the technical approach adopted in order to provide with an evolved solution in such a scenario. Finally, Sect. 5 summarizes the most notable contributions of this work, as well as future research avenues.

2 Related Work

Several approaches have been proposed to overcome the limitations of PACS systems. On a theoretical approach, Pohjonen et al. [5] revisited the grid computing and streaming concepts applied to a PACS system. Vossberg et al. [6] adapted the DICOM protocol to use a grid infrastructure, distributed across different institutions. Yang et al. [7] also used an analogous approach. Although they reportedly improved performance over other web PACS approaches, the grid infrastructure required a complex setup in both cases, and none of these works developed a DICOM viewer. Furthermore, data grids are usually employed when resources are required for solving large-scale, data-intensive scientific applications, which may be beyond medical imaging storage and access requirements in many cases.

Costa et al. [8] developed a PACS solution based on peer-to-peer (P2P) communication models, and utilized document-based indexing techniques within a DICOM network. The system is marketed by BMD Software, Lda. (Aveiro, Portugal). Although P2P may be a solution to increase the dynamics of DICOM nodes, or to reduce the latency, it may not be the most appropriate solution from a security point of view.

Works bridging mobile and PACS technologies are polarized into two distinct approaches with advantages and drawbacks to each. Early attempts to bridge portable devices and PACS technologies required implementing PACS functionality on the devices [9]. This approach was aimed at providing wireless PACS access within a given institution. This type of solution benefits from a seamless integration with the

PACS. However, it falls short when requiring external access. In such cases, web protocols, which are platform independent and can be leveraged to provide a common interface for multiple platforms, need to be used. Valente et al. [9] proposed a RESTful architecture, employing the well-known open source dcm4che library for DICOM communications. Although a medical imaging viewer was also proposed, their approach required some set-up, even if minimal, on the client side.

Finally, the advent of mobile devices provides flexibility of work, anytime, any-where. Although mobile environments have inherent constraints (such limited display size compared to traditional radiology viewers or limited computing capabilities), the FDA cleared in 2011 the first diagnostic radiology app [10] and the number of cleared apps has grown ever since.

The so-called zero-footprint, web-based DICOM viewers consist of applications that do not require end-users to install any software. Since only a browser is required and run on any platform, the solution is very cost-effective. Different commercial products [11, 12] offer such DICOM viewers. However, although some PACS solutions incorporating web-viewers claim to be zero-footprint solutions, yet many require plugins to run, only run on specific browsers, or do not support tablet or mobile device capabilities without further specialization. Hence, the true advantage of a zero-footprint, such as decreasing cost of ownership and deploying data with minimal or virtually no support, is diminished.

3 PACS Evolution

Transitioning to a new PACS vendor or platform has traditionally involved complex and expensive software and hardware migration efforts, caused by the proprietary mechanisms that have been at the core of such systems. Within recent years, alternative technologies have experienced an unprecedented development that has made the emergence of new actors in the medical imaging industry possible. At the same time, the complexity of the infrastructure required to develop and maintain medical image management systems has been reduced with respect to the amount of stored data. This has allowed novel services to be proposed to healthcare centers that would not have been possible otherwise.

To address these issues, the concept of Vendor Neutral Archives (VNA) [2] has emerged in the last few years. VNA solutions provide hospitals with a decoupled solution for image archiving, so traditional PACS have started to be shifted to these enterprise imaging solutions. With the introduction of VNA and cloud-based digital archives, data from all departments can be put into one pot and accessed using a universal method. This enables the data to be managed by the healthcare system's information technology (IT) staff, rather than individual departments or radiology units.

More recently, in late 2014, the notion of "deconstructed PACS" architecture was introduced as an evolution of VNA systems during the Radiological Society of North America (RSNA) meeting. The underlying idea behind a deconstructed PACS (also referred to as PACS 3.0), is not only decoupling the archive system, but also the diagnostic and clinical viewing of images, the enterprise workflow orchestration and the imaging analytics systems. Deconstructed PACS enable institutions to gain control of their images and optimize workflow, but perhaps most importantly, they can choose the best viewer available for interpretation and also the best analytical processes [13, 14].

4   Proposed Architecture

We have designed and developed an implementation for a deconstructed PACS relying on its hardware-agnosticism and modularity principles. Besides these features, other key aspects for its delivery "as-a-service" have been taken into consideration, such as the adoption of cloud-based solutions, data protection or cost-related issues. The sum of all these elements led our development into a new concept of "Deconstructed PACS-as-a-service" solution. The proposed architecture is depicted in Fig. 1.

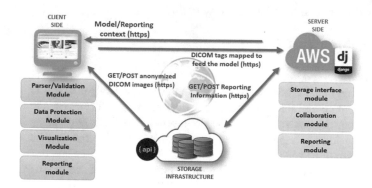

Fig. 1. Proposed architecture.

4.1   Hardware-Related Strategy

The proposed deconstructed PACS-as-a-service relies on a three-tier architecture: a client-side for end-user data access; a storage infrastructure as data repository and a server side to place the backend actions.

4.1.1   Client side
The developed interfaces have followed a responsive design, so that the application can be accessed from both PC or mobile devices. Although according to the IT Reference Guide for the Practicing Radiologist [15], higher resolution displays do not necessarily translate into better diagnostic quality, a set of minimal requirements related to aspects such as calibration or luminance are required to maximize diagnostic accuracy.

4.1.2   Storage infrastructure
All files managed by the application are stored into a cloud-based storage provider that can be selected by the user, provided that an API to communicate with third parties exists. With this strategy, all storage specific tasks are delegated to the cloud storage service, thus ensuring the right scalability of the system.

4.1.3   Server side

The system can be deployed into any "Platform-as-a-Service" provider. These kinds of services provide a platform that allows customers to develop, run and manage applications without the complexity of building and maintaining the infrastructure. Some of the common tasks they provide are capacity provisioning, load balancing or auto-scaling. We have successfully tested the deployment of our solution into an AWS Elastic Beanstalk environment.

4.2   Software-Related Strategy

The logic of the system is divided into the following modules: an acquisition module in charge of the image gathering process; a visualization module for image inspection; a searching module for locating tasks and finally a collaborative module where we have implemented several simple enterprise workflows.

4.2.1   Acquisition module

Parser/validation module
Digital Imaging and Communications in Medicine (DICOM) [16] is the standard for the communication and management of medical imaging information and related data. One of the goals of this standard is the definition of the Data Dictionary that shall be supplied alongside the images, to achieve a seamless data interchange between digital imaging computer systems.

We have integrated into our solution an adapted version of the dicomParser library [17] which is a lightweight tool for parsing DICOM byte streams in modern HTML5 based web browsers. The system discards non-DICOM-compliant files, and also verifies all mandatory DICOM fields.

Data protection module
The fast growth of digital media solutions has provided great advances in healthcare, but also has brought increased risk related with data protection. According to the European Parliament and Council 2016/679 Regulation of 27 April 2016 [18], data protection issues must be taken into consideration from the design phase of any digital solution. Pursuant to this legal resolution, we have implemented a data protection module in charge of anonymizing all sensitive DICOM tags. This module is based on a JavaScript cross-compilation of DCMTK [19], so it is executed on the client-side, minimizing the risks associated to personal data transfer to third-party cloud storage.

In addition to this data protection module, other security-related strategies have been also developed, such as the adoption of the HTTPS protocol, a 3-step registration design or a password expiration policy.

Storage interface module
We have developed a common interface capable of working with any public API exposed by cloud-based storage providers. The set of files sent to these storage providers comprise the DICOM images, the textual reporting information and the audios created for reporting (explained in more detail within the *Reporting module.* section).

This interface has been successfully tested against the Box Inc. [20] storage provider using the Box Python SDK [21].

4.2.2   Visualization module

We have implemented both 2D and 3D visualization (Fig. 2) modalities. Images are sent directly from the storage provider to the client side, without consuming resources from the backend server which helps to reduce costs and waiting times. Relevant DICOM metadata are presented alongside the image stacks. Furthermore, the system allows users to enrich this metadata with descriptors and comments.

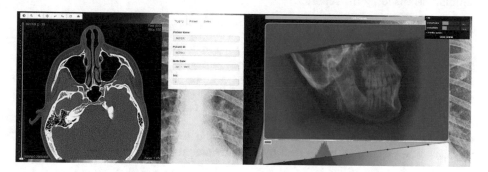

Fig. 2. Partial captures of the 2D (left) and 3D (right) visualizations of a dataset.

2D Visualization

We have integrated the Cornerstone JavaScript library [22] in order to provide 2D visualization capabilities in browsers that support the HTML5 canvas element. The Cornerstone WADO Image Loader [23] is used as the engine for retrieving DICOM images over HTTP. In addition, we have included alongside the display a set of common medical imaging interaction tools [24], such us window-level, pan, zoom or rotate. To allow a smoother navigation, datasets are delivered and rendered accordingly in cached blocks of 13 slices (this number was empirically set). Table 1 depicts caching and rendering times of one of these blocks with different resolutions.

Table 1. 2D visualization caching and rendering time measurements.

Resolution (size)	128 × 128 × 13 (0.5 MB)	256 × 256 × 13 (1.63 MB)	512 × 512 × 13 (6.53 MB)	1024 × 1024 × 13 (26 MB)
Cache/Render time	3510 ms	3755 ms	4660 ms	6919 ms

3D Visualization

Some applications rely on the ability of the viewer to offer 3D visualization for a better appreciation of patient anatomy and location of possible pathologies. For that purpose, we have included a web implementation of Direct Volume Rendering (DVR) [25, 26] using the X3DOM Library that offers implementation for most common representation use cases of volumetric medical imaging data [27]. Table 2 depicts mean preparation times for datasets with different resolutions before interactive rendering.

Table 2. 3D visualization rendering time measurements.

Resolution (size)	128 × 128 × 30 (1 MB)	256 × 256 × 13 (3.77 MB)	512 × 512 × 30 (15 MB)	1024 × 1024 × 30 (60 MB)
Render time	6301 ms	6681 ms	7740 ms	18202 ms

4.2.3 Search module

We have developed a retrieval tool for medical images capable of performing queries within a pre-defined set of fields including the name of the patient, the description of the study, the description of the series, the descriptors assigned to the images or the annotations made by the reporter.

4.2.4 Collaboration module

This module oversees ruling the three collaboration actions that users can perform with their images: (1) **Share**: users can share images with read-only permissions (2) **Request for reporting**: users can request a report for their cases. This action will emit a request for reporting to the recipients whom will be able to accept or reject the request (3) **Transfer ownership**: users can transfer the ownership of their series.

Reporting module. We have developed a specific module to manage the reporting requests. The main tasks performed by this module are:

- Manage the workflow of the request, providing state-aware information to both involved users (requestor and recipient).
- Provide with a set of 2D image interaction tools specific for reporting needs, such as Region of Interest (ROI) delineation, measurements (distances, angles), probe, annotation and screenshot.
- Provide users with three different, non-excluding ways of reporting cases:
  - RSNA templates: this module is able to communicate with the Radreport API [28] in order to assign to the requested case a "structured" report template which the recipient is expected to fill in. According to the RSNA Reporting Initiative, these templates: (1) Improve communication between radiologists and referring providers. (2) Enable radiology practices to meet accreditation criteria and (3) Help radiology practices earn pay-for-performance incentives.
  - Free text: recipients can report requested cases using free text. We have integrated the Web Speech API [29] to facilitate this task. This allows users to dictate the reports using a microphone; the Web Speech API automatically performs the speech-to-text process.
  - Audio files: cases can be reported using audio files. Once the speech has been recorded, the audio clips are available to be listened to within the application at a later time, using an embedded player.
- Export the reported information to a PDF file.

All these modules, as well as the wrapper that orchestrates them (the Python web framework 'Django') are based on open-source products, which facilitates the flexibility and maintainability of the solution.

5 Conclusions and Future Work

We have achieved to use existing web-based technologies like HTML5, JavaScript and WebGL to build a decoupled, scalable and secured tele-radiology solution based on extensively used cloud platforms like Amazon Web Services for web servers and Box for cloud storage. This work could define a new product concept called deconstructed PACS-as-a-Service to provide easy access of medical image content between institutions, their professionals and their patients.

According to the annual title "Imaging & Oncology" published by The Society of Radiographers [30], the adoption of this kind of solution would have a wide range of benefits for the 3 actors involved in a radiology process.

Patients can have immediate access to expert opinion, they benefit from enhanced speed and turnaround time of experts, from immediate availability of appropriate emergency investigations and immediate availability of a second opinion. Second, hospital management can supplement the lack of local expertise, it can face shortfalls of radiology provisions, balance demand variations, provide continuous service and availability of expert opinion. And finally, radiologists can review images from anywhere, they can manage more effectively their workload and provide their services across the world, along with the benefits of underlying stable working patterns and improved work life balance.

We have covered three aspects that typically compose this kind of hardware-agnostic solutions: image management and archiving, simple enterprise workflows orchestration and clinical viewing of images for diagnostic. Our work is modular and proves the maturity of existing technologies to support tasks like tele-radiology and remote collaborative exploration of medical images. Other aspects of deconstructed PACS approaches, like imaging analytics (along with all the necessary middleware), are being considered for future work. First, we will carry out a thorough study of more complex enterprise workflows that lead these kinds of processes in order to implement mechanisms to facilitate their integration. Secondly, we will work on improvements interoperability-related to our system, focusing on integrating information coming from the EHR using the HL7 FHIR (Fast Healthcare Interoperability Resources) [31] standard. DICOM tags will mapped to a FHIR resource instead of using an ad-hoc model. Another aspect to be considered is the fast deployment of automatic or semi-automatic image analysis processes (e.g. deep learning algorithms for detection, annotation, segmentation or screening) making use of actual technical products currently mature, to ship isolated processes fast (e.g. Docker containers). Finally, we will advance towards a unification of reporting criteria, implementing a comparison feature based on semantic analytics able to detect disparities between two different reports belonging to the same case.

Acknowledgments. This research work has been achieved in collaboration with the companies Bilbomática S.A., Derten Sistemas S.A. and Centro Médico Udalaitz S.A. It is part of the research project named "RADMOVE" (file reference # IG-2014/01264, IG-2015/00247, ZL-2016/00048) under HAZITEK support programs for the experimental development and industrial research projects of the government of the Basque Country (Spain).

References

1. Duerinckx, A.J.: Introduction to two PACS '82 panel discussions edited by André J. Duerinckx, M.D., Ph.D.: 'Equipment Manufacturers' view on PACS' and 'The Medical Community's View on PACS'. J. Digit. Imaging **16**(1), 29–31 (2003)
2. Agarwal, T.K., Sanjeev,: Vendor neutral archive in PACS. Indian J. Radiol. Imaging **22**(4), 242–245 (2012)
3. European Commission: Radiation Protection, Medical Radiation Exposure of the European Population, Part 1/2. Frost & Sullivan, Luxembourg, 180, November 2012
4. Frost & Sullivan: Prepare for Disasters & Tackle Terabytes When Evaluating Medical Image Archiving. Frost & Sullivan (2008)
5. Pohjonen, H., Ross, P., Blickman, J.G., Kamman, R.: Pervasive access to images and data - the use of computing grids and mobile/wireless devices across healthcare enterprises. IEEE Trans. Inf. Technol. Biomed. **11**(1), 81–86 (2007)
6. Vossberg, M., Tolxdorff, T., Krefting, D.: DICOM image communication in globus-based medical grids. IEEE Trans. Inf. Technol. Biomed. **12**(2), 145–153 (2008)
7. Yang, C.-T., Chen, C.-H., Yang, M.-F.: Implementation of a medical image file accessing system in co-allocation data grids. Future Gener. Comput. Syst. **26**(8), 1127–1140 (2010)
8. Costa, C., Ferreira, C., Bastião, L., Ribeiro, L., Silva, A., Oliveira, J.L.: Dicoogle - an open source peer-to-peer PACS. J. Digit. Imaging **24**(5), 848–856 (2010)
9. Valente, F., Viana-Ferreira, C., Costa, C., Oliveira, J.L.: A RESTful image gateway for multiple medical image repositories. IEEE Trans. Inf. Technol. Biomed. **16**(3), 356–364 (2012)
10. Press Announcements - FDA clears first diagnostic radiology application for mobile devices. http://www.fda.gov/NewsEvents/Newsroom/PressAnnouncements/ucm242295.htm. Accessed 06 Feb 2017
11. HTML5 Zero-footprint viewer for DICOM and PACS. LEAD Technologies, inc., 1927 South Tryon Street Suite 200 Charlotte, NC 28203 USA (2015)
12. Vue Motion: Carestream's Enterprise Access Viewer Connecting Technologies, Physicians, and Patients. Carestream Health, Inc. (2012)
13. Bassett, M.: PACS 3.0: the next iteration of radiology's reading platform. Radiol. Bus. J., 40–44, October/November 2016
14. Winthrop Resources: Finding ROI in Deconstructed PACS (2015)
15. Krupinski, E.A., Flynn, M.J.: IT Reference Guide for the Practicing Radiologist (2013)
16. DICOM Homepage. http://dicom.nema.org/. Accessed 30 Jan 2017
17. Hafey, C.: dicomParser - JavaScript parser for DICOM Part 10 data. GitHub. https://github.com/chafey/dicomParser. Accessed 30 Jan 2017
18. The European Parliament and the Council of the European Union: Regulation (EU) 2016/679 of the European Parliament and of the Council of 27 April 2016
19. dicom.offis.de - DICOM Software made by OFFIS - DCMTK - DICOM Toolkit. http://dicom.offis.de/dcmtk.php.en. Accessed 30 Jan 2017
20. Box Inc.: Box Content API. https://docs.box.com/docs. Accessed 30 Jan 2017
21. box/box-python-sdk. GitHub. https://github.com/box/box-python-sdk. Accessed 13 Mar 2017
22. chafey/cornerstone. GitHub. https://github.com/chafey/cornerstone. Accessed 30 Jan 2017
23. chafey/cornerstoneWADOImageLoader. GitHub. https://github.com/chafey/cornerstoneWADOImageLoader. Accessed 30 Jan 2017
24. chafey/cornerstoneTools. GitHub. https://github.com/chafey/cornerstoneTools. Accessed 30 Jan 2017

25. Arbelaiz, A., Moreno, A., Kabongo, L., García-Alonso, A.: Volume visualization tools for medical applications in ubiquitous platforms. In: eHealth 360°, pp. 443–450. Springer, Cham (2017)
26. Congote, J., Segura, A., Kabongo, L., Moreno, A., Posada, J., Ruiz, O.: Interactive visualization of volumetric data with WebGL in real-time. In: Proceedings of the 16th International Conference on 3D Web Technology, New York, NY, USA, pp. 137–146 (2011)
27. Arbelaiz, A., Moreno, A., Kabongo, L., García-Alonso, A.: X3DOM volume rendering component for web content developers. Multimed. Tools Appl., 1–30, July 2016
28. Developers | RadReport.org. http://radreport.org/dev/. Accessed 31 Jan 2017
29. Web Speech API Specification. https://dvcs.w3.org/hg/speech-api/raw-file/tip/speechapi.html. Accessed 31 Jan 2017
30. Dubbins, P.: Teleradiology – how much flatter is the world now? Imaging Oncol. **2009**, 56–59 (2009)
31. Index - FHIR v1.0.2. https://www.hl7.org/fhir/index.html. Accessed 31 Jan 2017

Heart Failure Readmission or Early Death Risk Factor Analysis: A Case Study in a Telemonitoring Program

Arkaitz Artetxe[1,2(✉)], Nekane Larburu[1], Nekane Murga[3],
Vanessa Escolar[3], and Manuel Graña[2]

[1] Vicomtech-IK4 Research Centre, Mikeletegi Pasealekua 57, 20009 San Sebastian, Spain
{aartetxe,nlarburu}@vicomtech.org
[2] Computation Intelligence Group, Basque University (UPV/EHU), P. Manuel Lardizabal 1,
20018 San Sebastian, Spain
manuel.grana@ehu.es
[3] Hospital Universitario de Basurto (Osakidetza Health Care System), Avda Montevideo 18,
48013 Bilbao, Spain
nekane.murga@gmail.com, vanessa.escolarperez@osakidetza.net

Abstract. Heart Failure (HF) is a clinical syndrome caused by a structural and/or functional cardiac abnormality that imposes tremendous burden on patients and on the healthcare systems worldwide. In this context, predictive models may facilitate the identification of patients at high risk of death or unplanned hospital readmissions and potentially enable direct specific interventions. Currently a plethora of studies in this field is discussing whether hospital readmission and mortality can be effectively predicted in patients with HF. In this work, we present a preliminary study for identifying risk factors for unplanned readmission or death, using a clinical dataset with 119 patients and 60 features. Different classification algorithms and feature selection approaches were employed in order to increase the prediction ability of the models and reduce their complexity in terms of number of features. Results show that sequential feature selection methods along with SVM achieve the best scores in terms of accuracy for predicting 30-day readmission or death risk.

Keywords: Heart Failure · Predictive models · Feature selection · Readmission risk

1 Introduction

Heart failure (HF) is a clinical syndrome characterized by typical symptoms (e.g. breathlessness, ankle swelling and fatigue) caused by a structural and/or functional cardiac abnormality, resulting in a reduced cardiac output and/or elevated intra-cardiac pressures at rest or during stress. Demonstration of an underlying cardiac cause is central to the diagnosis of HF. This is usually a myocardial abnormality causing ventricular dysfunction or abnormalities of the valves, pericardium, endocardium, heart rhythm and conduction [1].

The prevalence of HF is approximately 1–2% of the adult population in developed countries, rising to ≥10% among people >70 years of age [2].

© Springer International Publishing AG 2018
Y.-W. Chen et al. (eds.), *Innovation in Medicine and Healthcare 2017*, Smart Innovation,
Systems and Technologies 71, DOI 10.1007/978-3-319-59397-5_26

Over the last 30 years, improvements in treatments and their implementation have improved survival but the outcome often remains unsatisfactory. Most recent European data (ESC-HF[1] pilot study) demonstrates that 12-month mortality rates for HF patients are between 7% and 17%, and the 12-month hospitalization rates are between 32% and 44% [3].

The harms of cardiovascular disease (CVD) are not limited only to an individual's health. When CVD causes hospitalizations, short-term expenses tend to be extremely high. Costs include ambulance rides, diagnostic tests, hospital stays, and immediate treatment that may include surgery. Short-term costs aside, CVD remains expensive for the long-term due to the price of drugs, tests to monitor the progress of the disease, and frequent doctor appointments [4]. The high cost of CVD is compounded by the lack of productivity and income that such patient may have [5]. Additionally, high rates of readmission after hospitalization for HF impose tremendous burden on patients and on the healthcare system. In this context, predictive models facilitate identification of patients at high risk for hospital readmissions and potentially enable direct specific interventions toward those who might benefit most by identifying key risk factors. However, current predictive models using administrative and clinical data discriminate poorly on readmissions [17]. That is the reason why some studies have been developed in order to try to define whether machine learning would enhance prediction [5].

Nevertheless, it remains not clear whether we can predict and prevent hospital readmission and mortality in patients with HF. Currently, there are several programs where patient monitoring is carried out, so that clinicians can check patients' progress [7–11].

In some cases, clinicians define some simple rules, so that they can get some alerts that may indicate the deterioration of a patient [2]. In other cases, as shown in Mobiguide EU project, the system implements the local clinical guidelines and extend them to guide patients during their daily life [3]. But due to the lack of time of clinicians and lack of suitable IT solutions, clinicians do not exploit the monitored information.

The objective of this paper is twofold: First, we present a dataset built on a HF patient monitoring scenario. Second, we present a preliminary study for identifying risk factors associated to unplanned readmission or death.

The paper is organized as follows. In Sect. 2 we present some related works on predictive models and their impact in the healthcare domain. In Sect. 3 we present the dataset used for the study. Next, we describe the methods applied and the experimental results for each method. In Sect. 4 the results of each method are explained and compared in order to retrieve clinical conclusions. Finally, in Sect. 5 we discuss the conclusions and future work.

2   State of the Art

As presented in [6], predictive analytics can improve healthcare in several ways. For instance, they can be used to increase the accuracy of diagnosis, provide answers to specific cases of individual patients (not presented in clinical guidelines) or predict the

---

[1] Heart Failure – European Society of Cardiology.

medications that meet best the needs of smaller groups of patients. In the literature we found several studies that apply predictive models for different purposes.

Kansagara et al. [7, 8] presents a systematic review of 26 readmission risk prediction models of medical patients tested in a variety of setting and populations. Nevertheless, in their study they conclude that most readmission risk prediction remains poorly understood and has limitations. For example, factors such as social, environmental and access to care factors have not been widely studied.

In [5] it is shown that current predictive models using administrative and clinical data discriminate poorly on readmissions. The inclusion of a richer set of predictor variables encompassing patients' clinical, social, and demographic domains, while improving discrimination in some internally validated studies, does not necessarily markedly improve discrimination. Another possibility for improving models, rather than simply adding a richer set of predictors, is that prediction might improve with methods that better address the higher order interactions between the factors of risk. Many patients' risk may only be predicted by modelling complex relationships between independent variables.

3   Dataset Description

Since 2014 the OSI Bilbao-Basurto (Osakidetza), located in Basque Country, Spain, has a program to monitor heart failure patients. In this program the data is collected from diverse sources:

- *Baseline Information*: data that corresponds to the first seen by a physician when being diagnosed with HF. This information includes patient demographic information, such as year of birth and gender, but also clinical data, such as the hospitalization date, type of heart disease and hemodynamic parameters such as heart rate, systolic and diastolic blood pressure, blood checkup data, pharmaceutical treatment and other non-cardiac comorbidities.
- *Monitored Data*: data that is monitored by the patient remotely every week (with a frequency that varies from 3 to 7 days per week), which contains patient vital signs (such as heart rate, systolic/diastolic blood pressure, weight and oxygen saturation) and a questionnaire about the patient condition (e.g. *During the last 3 days, have you been having your medications as prescribed?*).
- *Patient Deterioration*: this contains the information that we aim to predict, which includes the readmissions, emergencies and mortality. For readmissions and emergencies, we get the detailed information about the type of readmission (e.g. if the readmission is related with HF) and the dates when the patient entered and left the hospital. For the mortality information, we also have the data and the causes (e.g. if the mortality is related with HF).
- *Alerts*: these alerts are either strong ("red") or moderate ("yellow") depending on the severity level and they are fired when previously defined conditions are fulfilled. They can be a 'simple rule' based on a single fact (e.g. *IF (heart rate < 55) THEN "yellow"*) or a 'tendency rule' based on the progress during certain time period of a vital sign (e.g. *IF (During 3 days weight loss 1 kg) THEN "red"*).

4   Methods

In this study our goal is to make a preliminary data analysis in order to figure out which features are more related to the HF readmission risk, using only baseline health status data shown in Table 1. The dataset is composed of 60 attributes collected from 119 patients with CVD from which 30 of the them were readmitted within 30 days (if a patient is readmitted more than once, only the first admission is included) and 12 died. With that in mind, we make use of feature subset selection techniques that allow us identifying the most significant variables or groups of variables of our dataset. In this

Table 1. Attributes of the dataset

Clinical history	*Laboratory*
Age (years)	Urea (mg/dl)
Gender	Creatinine (mg/dl)
Smoker (yes/no/former)	Sodium (mEq/L)
Weight (kg)	Potassium (mEq/L)
Height (cm)	Hemoglobin (g/dl)
HR – Heart Rate (bpm)	Total cholesterol (mg/dl)
SO2 – Oxygen saturation (%)	LDL cholesterol (mg/dl)
SBP – Systolic Blood Pressure (mmHg)	HDL cholesterol (mg/dl)
DBP – Diastolic Blood Pressure (mmHg)	Triglycerides (mg/dl)
Left Ventricular Ejection Fraction (%)	*Comorbidities*
Years since first diagnostic	Acute coronary syndrome
Admission days	Peripheral vascular disease
Implanted device	Stroke
Needs of oxygen	Dementia
Therapies	Chronic obstructive pulmonary disease
Furosemide	Connective tissue disease
Torasemide	Peptic ulcer disease
Thiazide	Mild liver disease
MRAs (Mineralocorticoid/aldosterone receptor	Diabetes mellitus
antagonists)	Hemiplegia
ACEIs (Angiotensin-converting enzyme inhibitors)	Moderate/severe renal disease
ARB (angiotensin receptor blocker)	Complicated Diabetes Mellitus
Beta blockers	Any tumor
Ivabrandine	Leukemia
Digoxin	Lymphoma
Anticoagulants	Moderate/severe liver disease
Antiplatelet therapy	Metastatic solid tumor
Oxygen therapy	Anxiety/depression
Antiarrhythmic drugs	Osteoarthritis/arthrosis/spondylitis
Lipid lowering therapy	Osteoporosis
	Sinus rhythm
	Atrial fibrillation
	Pacemaker rhythm

section we briefly present feature selection methods and the evaluation methodology that we followed to perform the experiments.

4.1 Feature Selection Methods

In supervised classification, a classifier is a prediction model built using a-training-dataset. The dataset is composed of a set of M instances, where each instance is described by a vector of features $X = (x_1, \ldots, x_n)$ and the class label $C = \{c_1, \ldots, c_n\}$. The classifier can be defined as a function g that returns the c value given a feature vector X (i.e. predicts the class of the input instance):

$$g:X \rightarrow C \tag{1}$$

$$g(x) = arg\ \max_c f(x, c) \tag{2}$$

where f defines a scoring function. The goal of feature subset selection is to find an optimal feature subset $X' \square X$ so that the accuracy of the classifier is maximal.

According to the taxonomy of feature selection techniques defined by Kohavi et al. [12] the methods can be grouped as follows: (i) Filter methods, (ii) Wrapper methods and (iii) Embedded methods. Following, the different techniques are briefly explained.

Filter Methods
According to [12] filter methods attempt to assess the merits of features from the data, ignoring the induction algorithm. A scoring function $S(i)$ is computed for each input variable x_i, (i[th] component of X) according to its corresponding c value.

Frequently, features are ranked according to their relevance, denoted by $S(i)$, assuming that high scores indicate high relevance and vice-versa. Eventually low-scoring features are removed, so that won't be eligible for further analysis or imputation to the classification algorithm.

In terms of computation, filter methods are efficient and scale well since they require only to compute n scores. However, its main advantage, that is, being classification algorithm agnostic, is at the same time one of its biggest disadvantages. The problem is that it ignores the effects of the selected feature subset on the performance of the classification algorithm. Another disadvantage that is usually pointed is that most proposed techniques are univariate [14]. It means that each feature is considered independently, without considering the interactions between different features. Not taking into account feature interactions can lead to model's suboptimal performance, inasmuch as features containing valuable information may score low and hence are not included in the model.

In order to overcome the problem of ignoring feature interactions, different multivariate techniques have been proposed. (e.g. correlation-based feature extraction [15]). Correlation-based feature selection (CFS) was the multivariate method that we utilized in our experiments.

Wrapper Methods
Unlike filter approaches, which ignore the biases of the classification algorithm, the wrapper approach makes use of a classifier for scoring the feature subset's predictive

power. As pointed in [12] the classifier is considered a black box, as no knowledge of the algorithm is needed, just the interface.

Wrapper methods conduct a search through the feature subset space for a good subset, where subsets are evaluated according to classifier's estimated accuracy. Classification model's accuracy is usually estimated using cross-validation.

Although in cases where the number of features is not too large an exhaustive search may be practicable, the problem is known to be NP-hard, what makes this approach computationally intractable [13]. Since an exhaustive search of the space is impractical, a search procedure guided by a heuristic function is defined. Multiple search strategies have been proposed, including hill-climbing, best-first or genetic algorithms among others.

One of the advantages of the wrapper approach is that interactions and dependencies between features are taken into account. Another advantage is that, unlike the filter approach, wrapper methods are linked to the classification model, so that the interactions of the feature set with the prediction model are considered. Nevertheless, a common drawback is that this approach is more prone to overfit to the training data. Wrapper methods are also criticized because their high computational cost, although efficient search strategies can alleviate the problem to a great extent.

In our experiments we have utilized Sequential Forward Selection (SFS) and Sequential Backward Selection (SBS) algorithms. Both are hill-climbing search algorithms that sequentially select features, until no improvement is observed in the evaluation function.

Embedded Methods
In these methods the search is conducted within the classifier itself, as part of the learning process. Embedded methods, in the same manner as wrapper methods, are tied to a specific classification algorithm. Nevertheless, the computational cost is significantly lower for embedded methods compared to wrapper methods and are less prone to overfitting than the latter.

In our experiments we have employed the well-known Random Forest (RF) algorithm, which is a tree-type embedded method.

4.2   Evaluation

In order to analyze which features are associated with HF readmission or death, we built classification models using different feature subsets. In this models, the outcome was the unplanned readmission or death within 30 days after discharge from HF hospitalization (0 for not readmitted, 1 for readmitted or dead). The evaluation of the models was made by performing 10 independent executions using leave-one-out accuracy estimation. Two well-known algorithms, namely Random Forest (RF) and Support Vector Machine (SVM) were used during the experiments.

5   Results

In this section we present the results obtained from the application of the following feature selection algorithms to our dataset:

- Correlation-based Feature Selection (CFS)
- Random Forest, embedded FS (RF)
- Sequential Forward Selection + SFS-SVM
- Sequential Backward Selection + SBS-SVM

As mentioned earlier, we used the well-known Random Forest (Gini as splitting criterion and 10 estimators) and SVM (radial basis function kernel, C = 1 and gamma = 1/number of features) classification algorithms, implemented in the open source machine learning library Scikit-learn.

Table 2 shows the mean accuracy along with the standard deviation of each model trained with the specified configuration. Results show that wrapper methods (using SVM) outperform other feature selection techniques. However, we observe that our models, regardless of the underlying method they utilize, perform poorly (below 67% accuracy).

Table 2. Mean accuracy and standard deviation for each classification algorithm and FS method

	None	CFS	RF	SFS-SVM	SBS-SVM
RF	.6227 ± .02	.6193 ± .03	.6353 ± .03	.6605 ± .02	.6454 ± .02
SVM	.6471 ± .00	.6471 ± .00	.6471 ± .00	.6639 ± .00	.6639 ± .00

Table 3 shows the list of features included by each method. For those randomized algorithms the number of times each feature was selected is shown. According to the results shown, several conclusions can be extracted:

- We observe that SBS method tends to be more homogeneous, since the majority of the selected features are present in multiple runs. It is noteworthy that 'years since first diagnostic' is a feature that is present in every execution, despite it is not present in the rest of methods. The reason may be related with the hill-climbing algorithm underlying, that is influenced by a local peak at the end part of the feature vector, so that features in this positions are more likely to be selected.
- On the other hand, SFS method selects a greater number of features although many of the selected features are only present in one of the runs.
- There is not a single feature that reaches the total consensus, that is, it is selected by all the methods at least in one run. Nevertheless, urea and pacemaker rhythm are two of the top features in terms of consensus, since they are present in all the FS method groups (i.e. filter, embedded and wrapper) and in many runs. (Figs. 1 and 2).

Table 3. List of variable included in the model by each method and number of times they were selected in the 10 randomized runs

Attribute	CFS	RF	SFS-SVM	SBS-SVM
Gender			1	
Smoker			1	
Weight	x	2	5	
Height		1		
HR		4		
SO2			2	
SBP		5		
Implant-dev			7	
Need oxygen			1	
Urea	x	10		7
Creatinine		4		
Sodium		1	1	
Potassium	x		1	
Hemoglobin			1	
Total cholesterol	x	1		2
HDL cholesterol		2	2	
Triglycerides	x	3		6
Torasemide	x		2	
Thiazide	x			
ACEIs			1	
ARB			3	
Ivabrandine	x			
COPD			1	
Connective tissue disease	x			
Peptic ulcer	x		4	
Diabetes mellitus			2	
Any tumor			1	
Moderate/severe liver disease				1
Metastatic solid tumor				1
Osteoarthritis/arthrosis/spondylitis				1
Osteoporosis	x			3
Sinus rhythm			1	
Atrial fibrillation				4
Pacemaker rhythm	x		8	7
Admission days		1	4	4
Age				2
Years first diagnostic				10

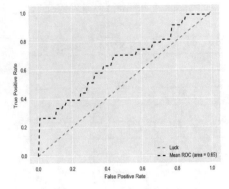

Fig. 1. SVM+SBS-SVM   **Fig. 2.** RF+SBS-SVM

6 Conclusions

In this work we have presented a preliminary study for identifying risk factors associated to unplanned readmission or death, using a HF clinical dataset. Different classification algorithms and feature selection methods were employed in order to increase the prediction ability of the models and reduce their complexity in terms of number of features. Results have shown that sequential (backward or forward) feature selection methods in combination with SVM perform the best in terms of estimated prediction accuracy. Nevertheless, according to the overall poor performance of the models, we hypothesize that baseline status data by itself may not have sufficient predictive capacity.

As future work we aim to study the monitored data to further improve the prediction of patient readmission or mortality. Additionally, we aim to develop a system that incorporates preventive actions. For that, we will develop a patient guidance system in a mobile platform, which will be based on the knowledge obtained from the predictive models, and the preventive actions that clinicians define.

Acknowledgement. This work was partially funded by Basque Government by means of the RIS3 Program.

References

1. Ponikowski, P., Voors, A.A., et al.: 2016 ESC guidelines for the diagnosis and treatment of acute and chronic heart failure. In: The Task Force for the Diagnosis and Treatment of Acute and Chronic Heart Failure of the European Society of Cardiology (ESC)
2. Mosterd, A., Hoes, A.W.: Clinical epidemiology of heart failure. Heart **93**, 1137–1146 (2007)
3. Ceia, F., Fonseca, C., Mota, T., Morais, H., Matias, F., De Sousa, A., Oliveira, A.G.: Prevalence of chronic heart failure in Southwestern Europe: the EPICA study. Eur. J. Heart Fail. **4**, 531–539 (2002)
4. Critical coverage for heart health: Medicaid and cardiovascular disease. American Heart & Stroke Association (2011). http://www.heart.org/idc/groups/heartpublic/@wcm/@adv/documents/downloadable/ucm_428187.pdf

5. Anand, S.S., Razak, F., Davis, A.D., Jacobs, R., Vuksan, V., Teo, K., Yusuf, S.: Social disadvantage and cardiovascular disease: development of an index and analysis of age, sex, and ethnicity effects. Int. J. Epidemiol. **35**(5), 1239–1245 (2006)
6. Mortazavi, B.J., et al.: Analysis of Machine Learning Techniques for Heart Failure Readmissions. doi:10.1161/CIRCOUTCOMES.116.003039
7. Riley, J.P., Cowie, M.R.: Telemonitoring in heart failure. Heart **95**(23), 1964–1968 (2009)
8. Inglis, S.: Structured telephone support or telemonitoring programmes for patients with chronic heart failure. J. Evid. Based Med. **3**(4), 228 (2010)
9. Cleland, J.G., Louis, A.A., Rigby, A.S., Janssens, U., Balk, A.H., Investigators, T.-H.: Noninvasive home telemonitoring for patients with heart failure at high risk of recurrent admission and death: The Trans-European Network-Home-Care Management System (TEN-HMS) study. J. Am. Coll. Cardiol. **45**(10), 1654–1664 (2005)
10. Lusignan, S., Wells, S., Johnson, P., Meredith, K., Leatham, E.: Compliance and effectiveness of 1 year's home telemonitoring. the report of a pilot study of patients with chronic heart failure. Eur. J. Heart Fail. **3**(6), 723–730 (2001)
11. United4Health. Transforming patient experience with telehealth in Europe. http://united4health.eu/. Accessed 3 Feb 2017
12. Kohavi, R., John, G.H.: Wrappers for feature subset selection. Artif. Intell. **97**(1), 273–324 (1997)
13. Guyon, I., Elisseeff, A.: An introduction to variable and feature selection. J. Mach. Learn. Res. **3**, 1157–1182 (2003)
14. Saeys, Y., Inza, I., Larrañaga, P.: A review of feature selection techniques in bioinformatics. Bioinformatics **23**(19), 2507–2517 (2007)
15. Hall, M.A.: Correlation-based feature selection for machine learning (Doctoral dissertation, The University of Waikato) (1999)
16. Watson, A.J., O'Rourke, J., Jethwani, K., Cami, A., Stern, T.A., Kvedar, J.C., Zai, A.H.: Linking electronic health record-extracted psychosocial data in real-time to risk of readmission for heart failure. Psychosomatics **52**(4), 319–327 (2011)
17. Kansagara, D., Englander, H., Salanitro, A., Kagen, D., Theobald, C., Freeman, M., Kripalani, S.: Risk prediction models for hospital readmission: a systematic review. JAMA **306**(15), 1688–1698 (2011)

Multimodal Imaging of the Breast to Retrieve the Reference State in the Absence of Gravity Using Finite Element Modeling

Remi Salmon[1](✉), Thanh Chau Nguyen[1,2], Linda W. Moore[3],
Barbara L. Bass[3], and Marc Garbey[1]

[1] Center for Computational Surgery, Houston Methodist Research Institute, Houston, TX, USA
rsalmon@houstonmethodist.org
[2] Center for Advanced Computing and Data Systems, University of Houston, Houston, TX, USA
[3] Department of Surgery, Houston Methodist, Houston, TX, USA

Abstract. We introduce in this work a new method of retrieving the reference state of the breast in a stress-free configuration and estimating at the same time the elasticity of the breast tissues by combining MRI and surface imaging data and using finite element analysis. This reference state of the breast is particularly useful in predicting the cosmetic outcome and the healing process of breast cancer surgery, and breast conserving therapy in particular.

Keywords: Breast cancer · Breast conserving therapy · Multiscale model · Finite element analysis · Reference state

1 Introduction

Breast cancer is the most prevalent form of cancer affecting women, accounting for 26% of cancer cases in the United States [1]. Additionally, breast cancer survival rate has been increasing during the past decades thanks to advances in early detections, mammography and general breast cancer awareness [2]. This increase in early detections and survival rate of breast cancer has resulted in the development of Breast Conserving Therapy (BCT) that combines localized breast cancer surgery with radiation and adjuvant therapy. By trying to preserve the contour of the breast, BCT improves the quality of life of the patient after surgery [3, 4], without impacting the cancer survival rate [5, 6]. The cosmetic outcome of BCT remains however less than optimal in many of the cases [7, 8].

In order to predict and help improve the cosmetic outcome of BCT we have developed a multiscale model that combines a finite element analysis of the deformation of the soft tissues of the breast after surgery at the macroscopic scale with a biological model of wound healing, operating at the spatial and time scale of a cell cycle. In our wound healing model, we assume that the breast is in a reference state defined as the breast geometry free of mechanical stresses, i.e. in the absence of gravity. We assume in other words that the topology of the cellular matrix can be modelized with a regular hexagonal grid in a fixed frame of reference that is not sensitive to the effect of gravity.

© Springer International Publishing AG 2018
Y.-W. Chen et al. (eds.), *Innovation in Medicine and Healthcare 2017*, Smart Innovation,
Systems and Technologies 71, DOI 10.1007/978-3-319-59397-5_27

We neglect as well the residual internal stresses resulting from the inversion of the gravity using finite element analysis. In this gravity-free state, the mechanical stress and strain present in the breast under the effect of gravity is taken into account in the probability of cellular division during the production of scar tissue [9, 10].

This reference state of the breast serves as a basis to our model by defining the "unloaded" frame of reference where the biological model of wound healing operates to model the closure of the wound over time. We also use this reference state to compute the contour of the breast and the stress distribution under the effect of gravity in a "loaded" frame of reference. The scheme of Fig. 1 summarizes these two frames of reference and how they interconnect in our multiscale model.

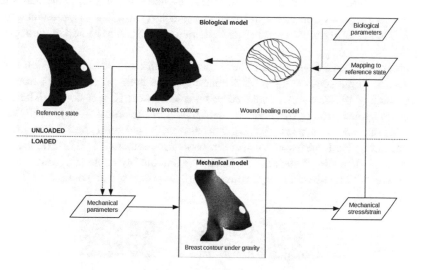

Fig. 1. Flowchart of the multiscale model of BCT, from [10]. The reference state of the breast (top left) is used as a basis to both define the gravity-free "unloaded" frame of reference where the healing model operates and to compute the effect of the gravity in the "loaded" frame of reference.

We have verified the feasibility of our model with a case study of a patient presenting an ideal tumor configuration in the center of the breast, where the reference state of the breast was retrieved from pre-operative MRI imaging only [10]. However the large variability in the anatomy of the patients enrolled in our clinical trial has shown the limitation of MRI imaging as a unique data source in our reconstruction of the reference state of the breast. Indeed, during the MRI acquisition the breast is often compressed onto the MRI coils, resulting in artifacts and missing information in the MRI data [11, 12]. We have also looked into using pre-operative surface imaging of the breast as input data to retrieve the reference state of the breast. However surface imaging of the breast fails to capture the whole breast contour when the data is acquired with the patients in a standing or sitting position. Indeed the breast is resting on the abdomen and the lower surface of the breast is not visible [13].

To account for this missing ground truth data in our model due to either (i) the compression of the breast on the MRI coils and (ii) the hidden surface of the breast when the patient is standing up, we present here a method to combine the pre-operative MRI imaging with the 3D surface imaging of the breast in order to improve the reconstruction of the reference state of the breast.

2   Methods

Determining the reference state of the breast is not a new issue, and previous studies by Rajagopal et al. have looked at retrieving the reference state of a finite element model of the breast using an iterative optimization algorithm [14]. The same technique has been used to perform multimodal image registration between MRI and mammography data [15].

In order to retrieve the correct, patient-specific geometry and elastic modulus of the breast, we combine pre-operative MRI imaging of the breast with a pre-operative 3D surface imaging of the breast, acquired using a Microsoft Kinect device. The reconstruction and generation of the finite element model of the breast from the MRI data is detailed in our previous work [9, 10]. The data collected with the Kinect device is processed using the RecFusion software in order to reconstruct a 3D surface mesh of the breast [16]. The Kinect surface reconstruction and the MRI data of a patient enrolled in our clinical trial is shown in Fig. 2; this data is then processed to isolate the 3D surface of the breast only.

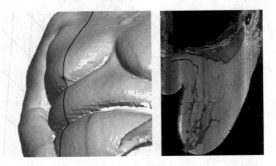

Fig. 2. Kinect Surface reconstruction of the patient acquired before surgery (left). The sagittal plane going through the nipple is drawn over the surface mesh (solid line). MRI image of the same patient in the same sagittal plane, acquired in the prone position (right).

The method, we propose here, makes use of multimodal imaging to retrieve both the missing boundary conditions in the reference state of the breast reconstructed from the pre-operative data as well as an estimation of the elastic modulus of the breast tissues. We show in the results section an application of our method on a 2D and a 3D model of the breast of the same patient enrolled in a clinical trial currently underway at the Houston Methodist hospital [17].

2.1 Two-Dimensional Simplification

We first consider in this study a 2D section of the breast in the sagittal plane going through the nipple for both the MRI and the Kinect surface data, assuming no displacement of the breast in the direction orthogonal to the sagittal plane. We inverse the effect of the gravity on the breast model by applying a body force opposite to the one of the gravity, see Fig. 3, using the finite element analysis software FEBio with no displacement boundary conditions on the top, bottom and back surfaces of the breast model [18]. As stated, we neglect here the residual internal stress in the breast after inversion of the gravity. We have developed on this subject an optimization algorithm to compensate for the residual internal stress detailed in Thanoon et al. [19]. We use a uniform hyperelastic Neo-Hookean material to model the breast tissues, parameterized in FEBio by its Young's modulus E and a density v. We fix the density in this study to $v = 0.49$, assuming the breast tissue be quasi-incompressible.

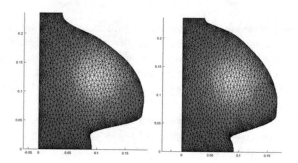

Fig. 3. 2D mesh of the breast reconstructed from the MRI data A (left), and result after inversion of the gravity A^* (right). Axis in meters.

Let us name $A = \{a_i\}$ the curve defining the skin envelope of the breast, i.e. the breast contour, retrieved from the MRI data acquired with the patient in the prone position, with $a_i = \{x_i, y_i\}, i \in [1, N]$. We note $A^* = \{a_i^*\}$ the breast contour A after inversion of the effect of gravity, where A^* is a function of the unknown Young's modulus E. Similarly, we name $B = \{b_i\}$ the skin envelope of the breast retrieved from the Kinect surface data, with $b_i = \{x_i, y_i\}, i \in [1, M]$.

Due to the fact that, for most patients enrolled in our clinical study, the breast of the patient is resting on the abdomen when the patient is standing up during the acquisition of the 3D surface imaging, as seen in Fig. 2, it is actually not possible to retrieve entirely the contour of the breast. To virtually recover this missing contour of the breast, we define the unknown lower extremity of the breast contour in the sagittal plane $b_B = \{x_B, y_B\}$ such that the curve going through $\{b_i, b_B\}$ is the contour of the entire breast. In practice, we define a sampling of the space of the 2D sagittal plane where b_B is a priori located, see Fig. 4. For each possible $b_B^k, k \in [1, K]$, where $K > 1$, we define a new curve $C^k = \{c_i\}$, interpolation of the parametric curve $\{b_i, b_B^k\}$ on a regular interval of $M^* > M$ points, with $c_i = \{x_i, y_i\}, i \in [1, M^*]$. The curve C^k is computed using the piecewise cubic Hermite interpolating polynomial function of MATLAB in order to ensure smoothness and C1 continuity. We fix $M^* = 150$ data points. The upper, lower and back sides used for the no displacement boundary conditions are created artificially in MATLAB in order to close the 2D breast contour. Finally, we name $C^{k*} = \{c_i^*\}$ the breast contour C^k after inversion of the gravity on the breast model.

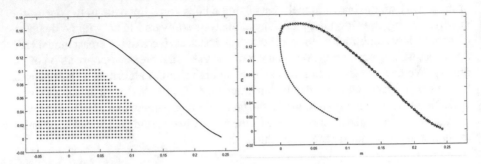

Fig. 4. Skin envelope of the breast B acquired from surface imaging (solid line, where the top of the breast is on the right end and the bottom on the left end) and sampling of the 2D space for the possible location of the k possible lower extremum points b_B^k (left). This particular grid of b_B^k points is chosen in order for the finite element simulation of the inversion of the gravity to converge without error from the solver. Sample of a generated 2D breast contour (right). Axis in meters.

In order to retrieve the unknown parameters of the model E and b_B^k, we minimize an objective function f defined as the average distance between the two breast contours in the absence of gravity A^* and C^{k*}:

$$f(E, k) = \frac{1}{M^*} \sum_i \left\| \tilde{a}_i^*(E) - c_i^*(E, k) \right\|$$ (1)

where $\tilde{a}_i^*$ is the interpolation of a_i^* on the regular interval defined previously for c_i^*. The minimum $\left(E_{min}, k_{min} \right)$ of the objective function is such that $A^*(E_{min})$ is the estimated reference state of the breast in our multiscale model. We show in Fig. 5 a diagram to illustrate the workflow leading to the evaluation of the objective function and the optimization of the parameters of our model.

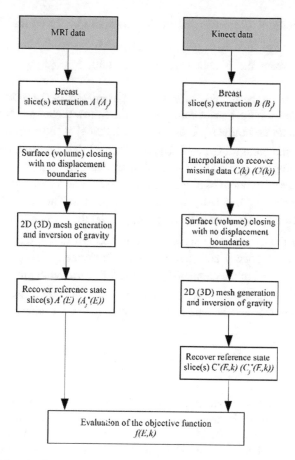

Fig. 5. Data workflow leading to the construction of the objective function f. The variable defined for the 3D formulation of the problem are noted between parentheses

2.2 Three-Dimensional Formulation of the Problem

We develop a similar method on a 3D mesh of the breast composed of a number S sagittal slices equally spaced for both the MRI and the Kinect surfaces of the breast. For those two datasets, each curve of the breast contour surface is closed using top, bottom and back surface boundaries identical to the ones shown in Fig. 3, with no displacement boundary conditions. From the resulting 3D cloud of points of the combined slices, we first generate a surface mesh using the Poisson surface reconstruction algorithm implemented in Meshlab [20]. Finally, a 3D volumetric mesh using tetrahedral volume elements is generated using the Gmsh software [21].

Following the same methodology, we name $A_j = \{a_i^j\}$ the collection of S slices defining the surface of the breast retrieved from the MRI data acquired with the patient in the prone position, with $a_i^j = \{x_i, y_i, z_i\}, i \in [1, N_j]$ and $j \in [1, S]$. We note

$A_j^* = \{a_i^{j*}\}$ the slice A_j after inversion of the effect of gravity on the 3D model of the breast. Similarly, we name $B_j = \{b_i^j\}$ collection of S slices defining the surface of the breast retrieved from the Kinect surface data, with $b_i' = \{x_i, y_i, z_i\}, i \in [1, M_j]$ and $j \in [1, S]$.

In order to reduce the number of unknown parameters in the 3D reconstruction of the breast surface from the Kinect data, we use the follow technique: we define the new unknown parameter point that $b_B^k, k \in [1, K]$ such that b_B^k is the projection of $b_{M_s}^k$ on the belly surface of the patient extracted from the Kinect data and interpolated under the surface of the breast, where s is the sagittal slice going through the nipple, see Fig. 2 (left). The degrees of freedom of b_B^k are then reduced to one by constraining the point b_B^k to belong to the interpolated belly surface of the patient. The curve L defined by the spline interpolation between the points $\{b_{M_1}^1, b_B^k, b_{M_s}^S\}$ is the lower boundary of the breast, initially missing from the Kinect surface imaging. Let us note l_k^j the intersection of the sagittal plane of the slice j and the curve L. We define a new collection of slices $C_j^k = \{c_i^j\}$, interpolation of the parametric curve $\{b_i^j, l_k^j\}$ on a regular interval of M^* points. Finally, the same algorithm for the evaluation of the objective function is applied; a diagram Fig. 5 summarizes the data workflow leading to the evaluation of the objective function f.

3   Results

We show here the result of our method on a 67 years old patient enrolled our clinical study, diagnosed with an invasive dual carcinoma on the lower inner quadrant of the right breast. Post-operative MRI and Kinect surface imaging were acquired at the Houston Methodist hospital. The MRI data was acquired with a voxel resolution of $07031 \times 0.7031 \times 1$ mm, and the Kinect surface data was acquired with an average resolution of 1.2 mm. For both data sets, a 2D triangular mesh and 3D tetrahedral mesh are generated with an average resolution of 2 mm.

In the 2D formulation of our problem, the objective function f defined in Eq. (1) was evaluated with a Young's modulus E ranging from 0.2 kPa to 7.5 kPa with a resolution of 0.1 kPa from 0.2 to 1 kPa and a resolution of 0.5 kPa from 1 to 7.5 kPa and a total number of 19 values; b_B was chosen on a grid of 596 points on a regular grid of 5 mm resolution, for a total of 19*596 simulation runs. The minimum of the objective function is obtained for $E_{min} = 0.6$ kPa, within the range of measured values for fat and glandular breast tissues in the literature [22]. We observe indeed on Fig. 2 (right) that the breast appears to have a large proportion of fat tissues that correlates with a lower Young's modulus.

The fitting between the breast contours reconstructed from MRI data and surface imaging of the breast after optimization of the objective function is shown in Fig. 6 (left). We observe a relatively good fit for the upper, "visible" section of the breast contour with an error within in the order of magnitude of the mesh resolution of 2 mm. The less optimal fit of the lower section of the breast can be explained by the use of interpolation where the surface imaging of the breast initially failed to capture the breast contour.

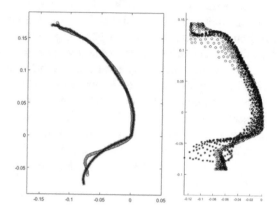

Fig. 6. Optimum fitting between the breast contour reconstructed from the MRI data (o) and the surface imaging of the breast (*) after minimization of the objective function f for the 2D model (left) and the 3D model (right) of the breast. We limit the nodes of the 3D finite elements model of the breast to the region within ± 4 cm of the sagittal plane going through the nipple to help the visualization and comparison with the 2D model of the breast. Axis in meters.

We then evaluated the objective function on the 3D breast contour, with a Young's modulus E ranging from 0.5 to 2 kPa with a resolution of 0.25 kPa; we then construct $k = 10$ different 3D breast model from the Kinect data corresponding to different values of b_B. An illustration of a subset of those models are shown in Fig. 7. We obtain a global minimum of the objective function f for $E = 0.75$ kPa, comparable to the result obtained with the 2D formulation of the problem (Fig. 7).

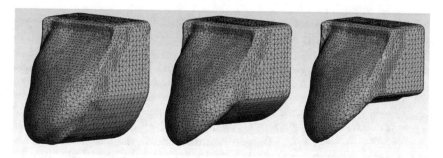

Fig. 7. Subset of the $k = 10$ 3D breast model reconstruction from the Kinect surface imaging of the breast obtained by varying the constraining the parametric point b_B^k, corresponding to different lower surface of the breast

4   Conclusion

We have detailed here a method making use of non-invasive, multimodal imaging of the breast of BCT patients that can successfully be used to accurately estimate simultaneously retrieve the breast elastic modulus, a critical parameter of our multiscale model, as well as the missing data of the breast contour when acquired with either MRI or surface imaging. This technique was easily extended to a three-dimensional analysis of the MRI and surface imaging data acquired in our clinical study with good results. We also plan to compare our estimation of the elastic modulus of the breast tissues with elastography measurements in order to complete the validation of our model.

The methods presented here provide a simple and non-invasive way to reconstruct a 3D, patient-specific reference model of the breast and will help further validate our model of the outcome of breast conserving therapy, in order to provide clinically relevant insights to both patients and surgeons.

Acknowledgements. The authors would like to thank the Department of Surgery at the Houston Methodist hospital and the patients enrolled in the clinical trial, the Horizon 2020 grant Desiree (Award# 690238) and all of its partners.

References

1. Jemal, A., Siegel, R., Ward, E., Hao, Y., Xu, J., Murray, T., Thun, M.J.: Cancer statistics, 2008. CA Cancer J. Clin. **58**(2), 71–96 (2008)
2. Ma, J., Jemal, A.: Breast cancer statistics. In: Ahmad, A. (ed.) Breast Cancer Metastasis and Drug Resistance, pp. 1–18. Springer, New York (2013)
3. Van Dongen, J.A., Bartelink, H., Fentiman, I.S., Lerut, T., Mignolet, F., Olthuis, G., Van der Schueren, E., Sylvester, R., Winter, J., Van Zijl, K.: Randomized clinical trial to assess the value of breast-conserving therapy in stage I and II breast cancer, EORTC 10801 trial. J. Natl. Cancer Inst. Monogr. **11**, 15–18 (1991)
4. Engel, J., Kerr, J., Schlesinger-Raab, A., Sauer, H., Hölzel, D.: Quality of life following breast-conserving therapy or mastectomy: results of a 5-year prospective study. Breast J. **10**(3), 223–231 (2004)
5. van Dongen, J.A., Voogd, A.C., Fentiman, I.S., Legrand, C., Sylvester, R.J., Tong, D., van der Schueren, E., Helle, P.A., van Zijl, K., Bartelink, H.: Long-term results of a randomized trial comparing breast-conserving therapy with mastectomy: European Organization for Research and Treatment of Cancer 10801 trial. J. Natl Cancer Inst. **92**(14), 1143–1150 (2000)
6. Morris, A.D., Morris, R.D., Wilson, J.F., White, J., Steinberg, S., Okunieff, P., Arriagada, R., Lê, M.G., Blichert-Toft, M., Van Dongen, J.A.: Breast-conserving therapy vs mastectomy in early-stage breast cancer: a meta-analysis of 10-year survival. Cancer J. Sci. Am. **3**, 6–12 (1996)
7. Clough, K.B., Cuminet, J., Fitoussi, A., Nos, C., Mosseri, V.: Cosmetic sequelae after conservative treatment for breast cancer: classification and results of surgical correction. Ann. Plast. Surg. **41**(5), 471–481 (1998)
8. Veiga, D.F., Veiga-Filho, J., Ribeiro, L.M., Archangelo-Junior, I., Balbino, P.F., Caetano, L.V., Novo, N.F., Ferreira, L.M.: Quality-of-life and self-esteem outcomes after oncoplastic breast-conserving surgery [Outcomes Article]. Plast. Reconstr. Surg. **125**(3), 811–817 (2010)

9. Garbey, M., Salmon, R., Thanoon, D., Bass, B.L.: Multiscale modeling and distributed computing to predict cosmesis outcome after a lumpectomy. J. Comput. Phys. **244**, 321–335 (2013)
10. Salmon, R., Garbey, M., Moore, L.W., Bass, B.L.: Interrogating a multifactorial model of breast conserving therapy with clinical data. PLoS ONE **10**(4), e0125006 (2015)
11. Yeh, E.D., Georgian-Smith, D., Raza, S., Bussolari, L., Pawlisz-Hoff, J., Birdwell, R.L.: Positioning in breast MR imaging to optimize image quality. Radiographics **34**(1), E1–E17 (2014)
12. Harvey, J.A., Hendrick, R.E., Coll, J.M., Nicholson, B.T., Burkholder, B.T., Cohen, M.A.: Breast MR imaging artifacts: how to recognize and fix them 1. Radiographics **27**((suppl_1)), S131–S145 (2007)
13. Lepoutre, N., Gilles, M., Salmon, R., Collet, C., Bass, B., Garbey, M.: A robust method and affordable system for the 3D-surface reconstruction of patient torso to evaluate cosmetic outcome after Breast Conservative Therapy. J. Comput. Surg. **1**(1), 11 (2014)
14. Rajagopal, V., Chung, J.H., Bullivant, D., Nielsen, P.M., Nash, M.P.: Determining the finite elasticity reference state from a loaded configuration. Int. J. Numer. Methods Eng. **72**(12), 1434–1451 (2007)
15. Rajagopal, V., Nash, M.P., Highnam, R.P., Nielsen, P.M.: The breast biomechanics reference state for multi-modal image analysis. In: International Workshop on Digital Mammography, pp. 385–392. Springer, Heidelberg, July 2008
16. Recfusion. http://recfusion.net/index.php/en/
17. Multi-scale Modeling of Breast Conserving Therapy (BCT). https://clinicaltrials.gov/ct2/show/NCT02310711
18. Maas, S.A., Ellis, B.J., Ateshian, G.A., Weiss, J.A.: FEBio: finite elements for biomechanics. J. Biomech. Eng. **134**(1), 011005 (2012)
19. Thanoon, D., Garbey, M., Kim, N.H., Bass, B.: A computational framework for breast surgery: application to breast conserving therapy. In: Garbey, M., Bass, B.L., Collet, C., Mathelin, M., Tran-Son-Tay, R. (eds.) Computational Surgery and Dual Training, pp. 249–266. Springer, New York (2010)
20. Cignoni, P., Callieri, M., Corsini, M., Dellepiane, M., Ganovelli, F., Ranzuglia, G.: Meshlab: an open-source mesh processing tool. In: Eurographics Italian Chapter Conference, pp. 129–136 (2008)
21. Geuzaine, C., Remacle, J.F.: Gmsh: a 3-D finite element mesh generator with built-in pre-and post-processing facilities. Int. J. Numer. Methods Eng. **79**(11), 1309–1331 (2009)
22. Rzymski, P., Skórzewska, A., Skibińska-Zielińska, M., Opala, T.: Factors influencing breast elasticity measured by the ultrasound Shear Wave elastography-preliminary results. Arch. Med. Sci. **7**(1), 127–133 (2011)

Short Papers

Applications for Mobile Devices and Methodologies for the Disorder Autism

Arnulfo Alanis Garza[1(✉)], Bogart Yail Marquez Lobato[1],
José Sergio Magdaleno Palencia[1], Maryuri Jacquelyne Chavez Duarte[2],
and Margarita Ramírez Ramirez[3]

[1] Computing and Systems Department, Technological Institute of Tijuana, Tijuana, BC, Mexico
{alanis,bogart,jmagdaleno,maryurid.chavez}@tectijuana.edu.mx
[2] Student-Computing and Systems Department, Technological Institute of Tijuana,
Tijuana, BC, Mexico
[3] Autonomous University of Baja California, Tijuana, BC, Mexico
maguiram@uabc.edu.mx

Abstract. This article summarizes some computer applications and methodologies for autism spectrum disorder and is presented to support intellectual and social evolution since this is a neurodevelopment disorders group affecting globally different higher brain functions of the individual, as intelligence, language ability and social interaction.

Keywords: Autism disorder · Software applications · Methodologies

1 Introduction

This article will explain several topics to discuss about autism reviewing what autism is as well as computer applications for both personal computer, and mobile devices.

Time constraints, only errors introduced during the preparation of the files will be corrected.

2 Autism Spectrum Disorder (Asd)

TEA diagnosis is often a two-stage process. The first stage involves an assessment of the overall development of a healthy child during visits with a pediatrician or health care provider in the early childhood. Children who show some developmental problems are channeled for further evaluation. The second step comprises a comprehensive evaluation conducted by a team of doctors and other health professionals with a wide range of specialties.

In this stage, a child may receive a diagnosis of autism or some other developmental disorder [1].

Y.-W. Chen et al. (eds.), *Innovation in Medicine and Healthcare 2017*, Smart Innovation,
Systems and Technologies 71, DOI 10.1007/978-3-319-59397-5_28

In general, a reliable autism spectrum disorder (ASD) diagnosis can be given to children of 2 years of age, although research suggests that some screening tests may be useful at 18 months

Many people-including pediatricians, family doctors, teachers and parents-may, at first, ignore the signs of ASD, believing that children will "reach" their peers. Although you might worry that your child has ASD, the earlier the disorder is diagnosed the faster interventions can begin. Early intervention can reduce or prevent the most severe disabilities associated with ASD. Early intervention can also improve the intelligence quotient (IQ) of your child, language and daily functional abilities, also called adaptive behavior [2].

3   History

The term autism first appears in the Dementia praecoxoder Gruppe der Schizophrenien 4 monograph drafted by Eugen Bleuler (1857–1939) the Treaty of Psychiatry directed by Gustav Aschaffenburg (1866–1944) and published in Vienna in 1911.

Bleuler replaces the notion of Dementia praecox disease that Emil Kraepelin (1856–1926) had define, based on a progressive evolution towards a terminal state of intellectual depletion (Verblodung) in a group of psychotic schizophrenics that had in common, whatever the clinical form under which manifest themselves such as a number of psychopathological mechanisms, the most characteristic the Spaltung (split) that gives its name to the group, and especially the core symptoms of autism. This term, created by Bleuler, has a Greek etymology "auto" meaning himself opposed to "another". Autism is characterized according to him by the withdrawal of the mental life of the subject about itself, reaching the constitution of a closed world separated from the external reality and the extreme difficulty or inability to communicate with others.

He wrote in 1911: a special and completely characteristic damage is that concerning the relation of the inner life to the outside world. The inner life takes a morbid predominance (autism). Autism is analogous to what Freud called auto-eroticism. But for Freud, eroticism and libido have a much more extensive significance than for other schools. Autism expresses the positive side of what Janet negatively named loss of sense of reality.

The sense of reality is not totally absent in schizophrenics. It lacks only for certain things that are contrary to their complexes. The Polish-born French psychiatrist Eugène Minkowski (1885–1972), an assistant for a while of Bleuler Burglözli in early World War and the introducer after the war of the phenomenological psychopathology in France, defined later autism within a perspective in reference to the notion of Élan vital introduced by the philosopher Henri Bergson, such as loss of contact with reality "Élan vital", a definition which will be the basis of his own conception of schizophrenia [3].

4   Implementation Types for ASD Detection

Sometimes the doctor will ask parents about the child's symptoms in order to detect the ASD. Other detection instruments combine information from parents with child observations made by the physician. Screening tools for infants and preschoolers examples include:

Verification Checklist for Autism in infants (CHAT, for its acronym in English) [4]

Modified Verification Checklist for Autism in infants (M-CHAT, for its acronym in English) [5].

Detection Tool of autism in children two years of age (STAT, for its acronym in English) [6].

Social communication questionnaire (SCQ, for its acronym in English) [7].

Communicative and symbolic behavior scales (CSBS, for its acronym in English) [8].

To detect mild ASD or Asperger's syndrome in older children, the doctor may depend on different detection instruments, such as:

- Autism spectrum scan questionnaire (ASSQ, for its acronym in English) [9].
- Australian Scale for Asperger syndrome (ASAS, for its acronym in English)
- Children Asperger Syndrome Test (CAST, for its acronym in English) [10].

Some useful resources for ASD detection include screening tools for the general development of the Center for Disease Control and Prevention and tools for specific detection of ASD on its website http://www.cdc.gov/ncbddd/Spanish/autism/screening.html.

5 Applications

Because of this problem with the disorder, be conducted a research and study how to help children with ASD, the approach is to investigate some of the applications, web and software that help support pages and you can have a better your quality of life.

There are many types of personal computer and smart mobile devices applications, whether Android or iOS platform that allow help children in the ASD.

Some of the applications are in games with the purpose of teaching some kind of linguistic and orthographic skills, simulation of communication such as chats (which are simulated environments), these are helpful to improve communication of children with this disorder.

Also new technologies can be an excellent support for a child with autism and the parents. For example, the Apple iPad has been considered especially useful for children with special needs and has been coming to homes, consultations and classes. Since it hit the market in April 2010 there have been more and more possibilities to use it as a support tool in special education. Teachers, parents and therapists describe that applications (Apps) and products from Apple and Android help develop different skills in children with autism.

For some children who cannot speak or have speech problems, iPad programs can be a support system for communication. Other applications and programs help children manage social situations that can be very stressful, like those that we've talked about agglomerations at hypermarket. Many other programs can help exercise fine motor skills like hand and finger movements quality control. These programs allow writing learning or better handling of delicate or small objects.

The advantage of these devices is that they are very simple and easy to handle: with just a touch something happens. They are very intuitive and attractive and children learn quickly to make good use of them.

Below are listed and described some of the applications that can be recommended:

- *Special words*

Teach children to recognize written words, using images and sounds. Includes 4 activities that increase in difficulty; matching pictures, words, and both. It can be customize by reordering, deleting or adding words, images and sounds. Supports, 1 or 2 Bluetooth buttons and content sharing via AirDrop, Dropbox, Google Drive and OneDrive.

- *Special stories*

Creating stories with text, images and sounds to later read and listen. Allows to print them via AirPrint, mail and share them via AirDrop, Dropbox, Google Drive, OneDrive. In iOS and OSX, allow to transfer them to iBooks.

- *Red cross*

Accident prevention and first aid. Teaches little ones to improve safety through accident prevention and learning the basics of first aid with 11 scenarios to simulate dangerous situations and their consequences.

- *ASD's doctor*

A website that tries to facilitate the most common medical visits for people with autism, familiarizing with the medical environment through a journey across different spaces, professional and medical procedures, which are explained in bullets, videos and animations.

- *Autism & PDD*

A set of 28 apps (many are free Litc versions) with various types of exercises based on stories and designed to develop language and also more than 10 other apps that teach appropriate behavior in school, home, etc.

- *Letters and I (Dyslexia)*

An interactive story to understand dyslexia adapted for iOS and Android, so that small kids understand what dyslexia is. Also helps parents to understand the difficulties of dyslexia. This comes in Castilian, Catalan, Basque, English and Italian.
Application available for Android and iOS.

- *Piruletras (Dyslexia)*

Children with dyslexia have difficulty learning to read and write. According to Luz Rello, Clara Bayarri and Azuki Gorriz, "helps children with dyslexia to overcome their reading and writing problems in Castilian through fun games". From reading and writing common errors among children with dyslexia they have been applied scientific methods to create games to help those children. There are three levels of difficulty with five types of exercises.

- *Follow me (Autism)*

For children with autism to evolve their visual attention and meaning acquisition. It is divided into six phases, from basic stimulation to meaning acquisition thru videos, photographs, drawings and pictograms. This app is aimed at children who do not yet read, cannot write, nor have understanding of the meaning of words and images, through games.

Free for iPad or Android tablets.

- *Special Words (Autism/Down syndrome)*

Special words, an app to recognize written words in order that children recognize written words, images and sounds thru four levels of difficulty with a total of 96 words, Also to capture the attention of children with Down syndrome, autism, hearing impairment and other learning difficulties. Is in Castilian and Catalan, among other languages.

App available for Android and iOS.

6   Methodologies

Here are some of the different methodologies, programs and techniques.

It is imperative that all specialized professional should be internalized in the diversity of methods, programs and techniques, to know the "what", "how", "when" and "who" to use them with.

Some are recognized as scientific, while others are known as alternative methodologies.

- *TCC-Cognitive-Behavioral Therapy*

Paradigmatic treatment, while new skills are developed through logical reasoning and persuasion motivation and with Socratic questioning (questions) challenging the irrational cognitions that maintain inappropriate behavior and skill practice by prioritizing in real situations (in life, in person, here and now).

- *ABA Methodology Applied Behavior Analysis*

The applied behavior analysis (Applied Behavior Analysis "ABA") was developed by Dr. Ivar Lova from the University of Los Angeles. It is a discipline dedicated to understanding and improving behavior, changing behaviors that affect learning.

7   Conclusions

Like these examples are many more. Today autism is more common than we think and many people have taken the task of making life easier through applications and software.

These applications are designed to facilitate emotional and intellectual development of children with ASD, and for their parents to develop their skills on how to educate them and also to provide better monitoring.

References

1. Filipek, P.A., Accardo, P.J., Ashwal, S., Baranek, G.T., Cook Jr., E.H., et al.: Practice parameter: screening and diagnosis of autism: report of the quality standards subcommittee of the american academy of neurology and the child neurology society. Neurology **55**(4), 468–479 (2000)
2. Bleuler, E.: Dementia praecox oder der gruppe der schizophrenien. Franz Deuticke, Leipzig y Wien (1991)
3. Minkowski, E.: La Schizophrénie, Nouvelle edn. Desclée de Brouwer, París (1927). 1953
4. Baron-Cohen, S., Allen, J., Gillberg, C.: Can autism be detected at 18 months? The needle, the haystack, and the CHAT. Br. J. Psychiatr. **161**(6), 839–843 (1992)
5. Robins, D.L., Fein, D., Barton, M.L., Green, J.A.: The modified checklist for autism in toddlers: an initial study investigating the early detection of autism and pervasive developmental disorders. J. Autism Dev. Disord. **31**, 131–144 (2001)
6. Stone, W.L., Coonrod, E.E., Turner, L.M., et al.: The screening tool for autism in two year olds can identify children at risk of autism. J. Autism Dev. Disord. **34**, 691–701 (2004). (STATPsychometric properties of the STAT for early autism screening)
7. Rutter, A.B., Lord, C.: SCQ: cuestionario de comunicación social: manual (2005). Wetherby, A.M., Prizant, B.M.: CSBS Cuestionario del bebé y del niño y pequeño (2001)
8. Wetherby, A., Prizant, B.: Communication and Symbolic Behavior Scales Developmental Profile- Preliminary, Normed edn. Paul H. Brookes Publishing Co., Baltimore (2002)
9. Ehlers, S., Gillberg, C., Wing, L.: A screening questionnaire for Asperger syndrome and other high-functioning autism spectrum disorders in school age children. J. Autism Dev. Disord. **29**(2), 129–141 (1999). PMID: 10382133
10. Scott, F.J., Baron-Cohen, S., Bolton, P., Brayne, C.: The CAST (Childhood Asperger Syndrome Test): Preliminary development of a UK screen for mainstream primary-school-age children. (http://www.autismresearchcentre.com/docs/papers/2002_Scott_etal_CAST.pdf)

Training Simulator for Resuscitation of Neonate with High Effectiveness and Low Introduction Cost

Noboru Nishimoto[1(✉)], Wei Yaguang[1], Kohei Matsumura[1], Roberto Lopez-Gulliver[1], Haruo Noma[1], Iwanaga Kogoro[2], and Tomohiro Kuroda[3]

[1] Ritsumeikan University College of Information Science and Engineering, Kyoto, Japan
nnishimoto@mxdlab.net
[2] Department of Pediatrics, Kyoto University Hospital, Kyoto, Japan
[3] Division of Medical Information Technology and Administration Planning,
Kyoto University Hospital, Kyoto, Japan

Abstract. The Japanese Society of Perinatal and Neonatal Medicine established the Neonatal Cardio Pulmonary Resuscitation (NCPR) workshop in 2007. The goal of the NCPR workshop is to train all medical staff to learn the algorithm to resuscitate neonates at all times after delivery. We designed a new training simulator for neonatal resuscitation that is both highly effective and has a low introduction cost.

Keywords: Neonatal simulator · Awareness · Video debriefing

1 Introduction

About 1 million babies were born in Japan in 2015. Most of them were in good condition, but more than 100,000 newborns required some treatments to stabilize their breathing just after delivery. The Neonatal Cardio Pulmonary Resuscitation (NCPR) workshop was started in 2007. The goal of the NCPR workshop is to train all medical staff to learn the algorithm to resuscitate neonates at all times after delivery. In the actual training, trainees need to be aware of the baby's condition by using a neonatal simulator to practice the resuscitation scenario to understand the algorithm. In this paper, we report our newly designed a training simulator for neonatal resuscitation that is both highly effective and inexpensive to introduce it.

Various training simulators have been used in the training. These can be categorized into high-performance simulators and simple simulators[1] The high-performance ones can simulate not only a baby's shape but also vital reactions such as breathing, heartbeat, temperature, and crying. However, such simulators are extremely complicated and expensive (more than twenty thousand USD) that they are not widely used. In contrast, the simple ones simulate a baby's shape and weight only (Fig. 1a), so they are easy and

---

[1] Sakamoto Model Copr., Sakamoto Baby Touch, http://www.sakamoto-model.co.jp/product/physical/m179/

© Springer International Publishing AG 2018
Y.-W. Chen et al. (eds.), *Innovation in Medicine and Healthcare 2017*, Smart Innovation,
Systems and Technologies 71, DOI 10.1007/978-3-319-59397-5_29

cheap (less than two hundred USD) to introduce widely[2]. The drawback is that these models cannot simulate any vital signs. Thus an instructor informs the trainees of such vital signs in different ways, such as by oral instruction or by tapping the desk with the hand (Fig. 1b). This results in decreasing the realism of the training. Additionally, the instructor has to perform many tasks simultaneously, for example, give instructions, control the training scenario, control the simulator, observe and record the trainees' actions following a paper checklist for debriefing after training. Thus instructors are required to be highly skilled.

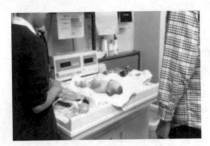

a) Baby simulator model   b) Instructor mimicking heartbeat by tapping the desk

Fig. 1. Current resuscitation scenario training using a simple simulator.

2   Proposed System

When designing our simulator, we ensure the following three requirements: (1) a subjective learning style for trainees, (2) an efficient lecture structure without requiring highly skilled instructors, and (3) a simple structure that leads to a low introduction cost. A block diagram of the proposed system is shown in Fig. 2.

First, in an actual auscultation, medical staff does not listen to any sound without first placing the stethoscope over the patient's body. When we simulate this action in the training, it is necessary to detect when the stethoscope is placed over the neonatal model. Inside the stethoscope we placed a light sensor that detects the light change when the stethoscope is placed on the baby model body. This light sensor is connected to a sound play module, that triggers a vital sound when trainees perform auscultation. We also developed a virtual pulse oximeter embedded into a micro-computer. These are connected to a PC-based simulator controller via a wireless connection.

Second, to make the instructor's task less complicated, the system allows the instructor to change a simulated vital sign simply by pressing the plus or minus button on a typical fame controller (Fig. 3c). Additionally, instead of reviewing the paper checklist, the trainees debrief their activity via a video that is recorded by the system during training. The instructor bookmarks the video by simply pressing the 'A' button

---

[2] American academy of Pediatrics, NeoNatalie Newborn Educational Mannequin https://shop.aap.org/neonatalie-newborn-educational-mannequin-dark-skin-kit/.

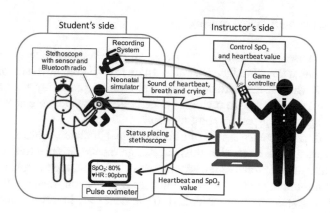

Fig. 2. Proposed system

when she/he notices any point that requires attention later. During the debriefing, the instructor effectively plays back bookmarked points on the video using the tags jump function (Fig. 3d).

Finally, our trial system is composed of commonly available electric parts and a standard PC, which makes it relatively inexpensive.

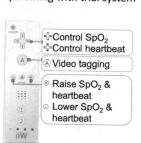

a) Training with trial system

b) Instructor handling the game controller

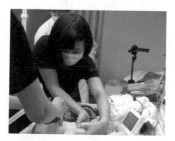

c) Controller interface

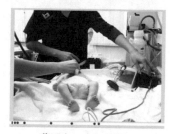

d) Video debriefing

Fig. 3. Resuscitation scenario training using the trial system.

3   Discussion and Improvements

We performed a user study with one instructor and two trainees to test our design. The trainees commented that "I judged the situation by myself and moved to the next action easily" and "When I listened to the heartbeat, I could judge my next action". The instructor stated that the "video review is effective because we were able to look back at the movie from a third-person point of view".

The current trial system consists of a simple baby simulator, a custom-made stethoscope, a PC, and a game controller, thus our design meets our objectives of offering an effective training at a reasonable introduction cost. It is possible to replace the PC with a tablet, so at minimum the system would just need the custom-made stethoscope. This means that our system can be distributed to medical staff in developing countries at reasonable cost.

References

1. Nehring, W.M.: US boards of nursing and the use of high-fidelity patient simulators in nursing education. J. Prof. Nurs. **24**, 109–117 (2008)
2. McNeal, G.J.: Simulation and nursing education. ABNF J. **22**, 284–292 (2010)

Effects of Depth Cues on the Recognition of the Spatial Position of a 3D Object in Transparent Stereoscopic Visualization

Yurina Kitaura[1]([⊠]), Kyoko Hasegawa[2], Yuichi Sakano[3,4], Roberto Lopez-Gulliver[2], Liang Li[2], Hiroshi Ando[3,4], and Satoshi Tanaka[2]

[1] Graduate School of Information Science and Engineering,
Ritsumeikan University, 1-1-1 Noji-higashi, Kusatsu, Shiga, Japan
is0163kk@ed.ritsumei.ac.jp
[2] College of Information Science and Engineering,
Ritsumeikan University, 1-1-1 Noji-higashi, Kusatsu, Shiga, Japan
{hasegawa,gulliver,stanaka}@media.ritsumei.ac.jp,
liliang@fc.ritsumei.ac.jp
[3] Center for Information and Neural Networks (CiNet),
National Institute of Information and Communications Technology,
3-5 Hikaridai, Seika-cho, Kyoto, Japan
{yuichi,h-ando}@nict.go.jp
[4] CiNet and Graduate School of Frontier Biosciences, Osaka University,
1-4 Yamadaoka, Suita, Osaka, Japan

Abstract. Medical applications, as well as many other scientific fields, frequently utilize transparent viewing to investigate the inner 3D structures of complex objects. On the other hand, it is known that stereoscopic vision is effective in allowing us to intuitively understand 3D shapes and to realize natural depth feel of visualized scenes. It is expected that the combination of these two visualization techniques, that is, transparent viewing and the stereoscopic vision, namely transparent stereoscopic visualization, should be effective for our easier and intuitive understanding of inner structures of 3D objects. However, the cognitive effects that arise when combining these two techniques have not been fully understood for us until now. In this paper, we investigate the cognitive effects that arise when combining these two techniques of computer visualization. We specially focus on medical volume visualization to investigate influences of the luminance gradient, which is inherent in the stochastic point-based rendering (SPBR) that we proposed recently. We conducted psychophysical experiments in which observers analysed the perceived 3D structure based on transparent stereoscopic visualization. The experiments are executed under the conditions of monocular, binocular viewing and motion parallax. We found that the luminance gradient is effective in the perceived depth magnitude in the transparent stereoscopic viewing of medical volume data.

Keywords: Transparent visualization · Automultiscopic 3D image · Depth perception

© Springer International Publishing AG 2018
Y.-W. Chen et al. (eds.), *Innovation in Medicine and Healthcare 2017*, Smart Innovation, Systems and Technologies 71, DOI 10.1007/978-3-319-59397-5_30

1 Introduction

Medical applications, as well as many other scientific fields, frequently utilize transparent viewing which make it possible to visualize not only the frontal surfaces of an object but also its complex 3D internal structure, otherwise invisible when using opaque viewing. Although it is important for the visualized translucent objects to be perceived as they physically are in terms of 3D structure, little has been known whether and in under which conditions stereoscopic vision is effective while visualizing the translucent objects. Other studies have been investigate perception of the opacity object and reported that higher luminance contrast between an object and the background causes an object to appear to be closer to the observer [1–3]. However, further investigation is needed in the case of translucent objects, because luminance value and luminance contrast are affected by the object's opacity.

In this paper, we investigate the effectiveness of combining transparent viewing with stereoscopic vision to visualize medical volume data. We conducted psychophysical experiments in which observers reported the perceived 3D structure using a five-view auto-multiscopic 3D display. As the transparent viewing method, we used the stochastic point-based rendering (SPBR) [4–6], which is known to be an effective and precise transparent rendering method. SPBR has its inherent shading effect, which we call "luminance gradient" in this paper. We investigate the influence of the effects in the perceived depth magnitude using the combination of transparent stereoscopic visualization.

2 Stochastic Point-Based Rendering and Luminance Gradient

The SPBR is a method of transparent rendering that uses small opaque particles as rendering primitives, which can create transparent images with the correct depth feel without the need for particle sorting. Additionally, this method relies on various types of fused visualization, such as surface-surface fusion and surface-volume fusion, to create the correct depth feel.

To render a surface, this method uses three processes. First, we generate particles based on L_R and opacity α as input parameters. Second, particles are divided into L_R groups and projected onto the image plane, so that L_R images are generated. Third, L_R images are averaged at the pixel scale to produce the resulting image. Thus, L_R performs rendering effectively.

The opacity α of a surface is based on the particle density. In other words, the higher the number of particles N is, the higher the generated opacity of the surface.

The relationship between these variables is described as follows:

$$n = \frac{\ln\left(1 - \alpha\right)}{\ln\left(1 - s/S\right)} L_R, \tag{1}$$

where s is the cross-sectional area of particles and S is total area of the surface. If the angle between the normal vector of the surface and the line of sight increases, the virtual opacity of the surface also increases. Thus, this relationship produces a luminance gradient on a smoothly curved surface. In this paper, we eliminate this property to examine the associated effect. Therefore, we modified the equation above as follows:

$$n = \frac{\ln(1-\alpha)}{\ln(1 - s/S|\cos\theta|)} L_R,$$

(2)

where θ is the angle between the normal vector of the surface and the line of sight.

3 Psychophysical Experiment

3.1 Methods

Eleven subjects were invited to participate in the experiment. They all took a visual stereo fly test and had normal or corrected-to-normal vision. We used a TRIDELITY MV4200, which is a 42-inch five-view auto-multiscopic 3D display with a parallax barrier (1920 × 1080 pixels), for the test. The viewing distance for the best 3D image quality of the display was 350 cm. In the experiment, the subjects viewed the display at this distance.

Based on various stimuli, CT data from the head were collected as volume data, and the isosurfaces of a simulated heavy particle irradiation dose were collected as surface data. These data are rendered using the SPBR method. The isosurface data included opacities of 0.05, 0.08, 0.1, 0.2, 0.3, 0.4, 0.5, 0.6, 0.7, 0.8, 0.9, and 0.99 rendered with or without a luminance gradient (Fig. 1).

The subjects viewed the stimuli based on four conditions: (a) both binocular disparity and motion parallax were available, (b) only motion parallax was available, (c) only binocular disparity was available, and (d) neither binocular disparity nor motion parallax was available. In the motion parallax condition, the subjects moved their heads laterally so that they observed all the five-view images; otherwise, head motion was restricted with a chin rest. In the binocular disparity condition, the subjects viewed the stimuli binocularly; otherwise, one of the subject's eyes was occluded by an opaque eye patch.

The experiment was conducted in a normally bright room. During the experiment, the subjects remained seated on a chair. The subjects were asked to report the depth magnitude of the isosurface in written, giving a number based on the assumption that the length of one side of a green square (upper right in Fig. 2) was 1.0. The ground truth value (i.e., correct answer) was 7.1. The magnitude of binocular disparity of the isosurface was 3.2 arcmin, assuming the inter-ocular distance was 6.0 cm. There were 96 conditions (4 viewing conditions, 12 opacities, and 2 luminance gradients). Each subject reported the depth magnitude of the isosurface once for each of all the conditions. The order of the conditions (viewing condition, opacity, and luminance gradient) was randomized.

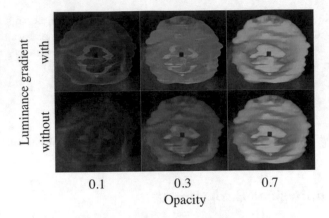

Fig. 1. Example of the isosurface

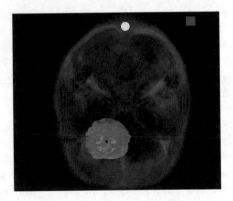

Fig. 2. Stimuli used in the experiment

3.2   Results and Discussion

Figure 3 shows that a high opacity of the isosurface led to a large perceived depth magnitude as reported by the subjects. In other words, high opacity is effective for accuracy of perceived depth. This result potentially occurs because the isosurface shading is weak to correctly perceive the structure or because large luminance contrast between the isosurface and the background at high opacity causes forefront of the isosurface to appear to be closer to the subjects. Additionally, introducing the luminance gradient inherent in the SPBR method caused the perceived depth magnitude of the isosurface to increase when the opacity was low. This result might be due to increase in luminance contrast between the isosurface and the black background. It has been reported that higher luminance contrast causes an object to appear to be closer to the observer [1–3, 7].

The perceived depth magnitude improved by introducing binocular disparity, motion parallax or both, compared to neither binocular disparity nor motion parallax (Fig. 3(a), (b), (c) vs. (d)). Other studies have reported that binocular disparity and motion parallax can have substantial effects on perceived depth [8–10].

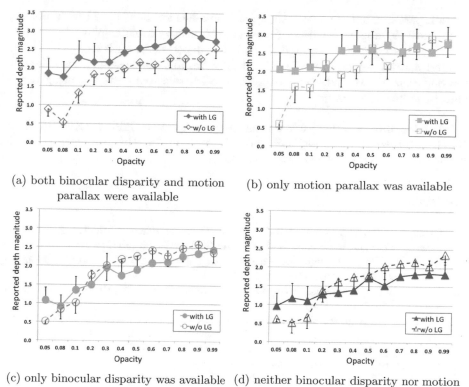

(a) both binocular disparity and motion parallax were available

(b) only motion parallax was available

(c) only binocular disparity was available

(d) neither binocular disparity nor motion parallax was available

Fig. 3. The perceived depth magnitude of the isosurface under various conditions. Error bars indicate SEM. LG: luminance gradient inherent in the SPBR method.

However, all perceived depth magnitudes were smaller than the ground truth value. In this case, subjects may have had difficulty assessing the measured depth magnitude because the gradient was too high.

4   Conclusions

In this research, we investigated the effects of depth cues in transparent stereoscopic visualization. According to our experiments, the perceived depth magnitudes become larger by introducing the luminance gradient, especially in low opacity regions. This means human depth perception, which is impaired in the transparent viewing, can be partially compensated by the luminance gradient. Note that the shape visualized in the experiments is a concavo-convex shape, for which effects of the luminance gradient become apparent. The luminance gradient should work effectively especially for transparent viewing of such shapes. It is also interesting that effects of the luminance gradient become stronger along with motion parallax. This suggests that the luminance gradient would be important in medical visualization with a glasses-free 3D stereoscopic display.

Further studies are required using other depth cues, such as colour, background and binocular disparity. Eventually, we aim to develop a visualization method that allows observers to perceive 3D structures correctly using depth cues in an effective manner.

Acknowledgement. This work was supported in part by JSPS KAKENHI Grant Number 16H02826 and MEXT-Supported Program for the Strategic Research Foundation at Private Universities (2013–2017).

References

1. Farne, M.: Brightness as an indicator to distance: relative brightness per se or contrast with the background? Perception **6**(3), 287–293 (1977)
2. Egusa, H.: Effect of brightness on perceived distance as a figure-ground phenomenon. Perception **11**(6), 671–676 (1982)
3. O'Shea, R.P., Blackburn, S.G., Ono, H.: Contrast as a depth cue. Vision Res. **34**(12), 1595–1604 (1994)
4. Tanaka, S., Hasegawa, K., Shimokubo, Y., Kaneko, T., Kawamura, T., Nakata, S., Ojima, S., Sakamoto, N., Tanaka, H.T., Koyamada, K.: Particle-based transparent rendering of implicit surfaces and its application to fused visualization. In: Euro Vis 2012, pp. 25–29 (2012). (short paper)
5. Hasegawa, K., Ojima, S., Shimokubo, Y., Nakata, S., Hachimura, K., Tanaka, S.: Particle-based transparent fused visualization applied to medical volume data. Int. J. Model. Simul. Sci. Comput. **4**, 11 (2013). 1341003
6. Tanaka, S., Hasegawa, K., Okamoto, N., Umegaki, R., Wang, S., Uemura, M., Okamoto, A., Koyamada, K.: See-through imaging of laser-scanned 3D cultural heritage objects based on stochastic rendering of large-scale point clouds. In: ISPRS Annals of Photogrammetry, Remote Sensing and Spatial Information Sciences, vol. III-5, pp. 73–80, July 2016
7. Sakano, Y., Kitaura, Y., Hasegawa, K., Lopez-Gulliver, R., Ando, H., Tanaka, S.: Evaluation of perceived 3D structure of multi-view 3D medical image based on transparent visualization: a psychophysical study. In: Proceedings of the International Display Workshops, vol. 23, pp. 898–901 (2016)
8. Howard, I.P., Rogers, B.J.: Binocular Vision and Stereopsis. Oxford University Press, New York (1995)
9. Julesz, B.: Binocular depth Perception of computer-generated patterns. Bell Syst. Tech. J. **39**, 1125–1162 (1960)
10. Rogers, B., Graham, M.: Motion parallax as an independent cue for depth perception. Perception **8**, 125–134 (1979)

A Quantitative Study of Local Ternary Patterns for Risk Assessment in Mammography

Andrik Rampun[1]([⊠]), Philip J. Morrow[1], Bryan W. Scotney[1],
and John Winder[2]

[1] School of Computing and Information Engineering,
Ulster University, Coleraine BT52 1SA, Northern Ireland
{y.rampun,pj.morrow,bw.scotney}@ulster.ac.uk
[2] School of Health Sciences, Institute of Nursing and Health,
Ulster University, Newtownabbey BT37 0QB, Northern Ireland
rj.winder@ulster.ac.uk

Abstract. This paper presents a preliminary quantitative study for breast cancer risk assessment in mammography using mathematical operators called Local Ternary Patterns. The study covers three different mapping patterns namely uniform ('u2'), nonuniform ('ri') and a combination of uniform and nonuniform ('riu2'). These patterns are used as texture features to model the appearance of breast density within the fibroglandular disk area. Subsequently, the Support Vector Machine is employed as a classification approach and initial results suggest that the mapping pattern 'riu2' outperforms the others.

Keywords: Computer aided diagnosis · Local ternary patterns · Breast cancer and mammography

1 Introduction

The use of Computer Aided Diagnosis (CAD) systems as a second reader opinion is increasingly popular in medical applications as it helps radiologists to interpret images efficiently and to produce results consistently. One of the elements that determines the reliability of a CAD system is the use of robust descriptors to capture distinctive characteristics of the region of interest in an image. Many texture descriptors have been investigated in the literature for breast density classification in mammograms. The most popular are based on the first and second-order statistical features due to its simplicity. Texture descriptors such as local binary patterns (LBP) and textons are also quite popular due to their ability to capture rich and descriptive characteristics of the texture. We refer the reader to the study in [1] for full reviews of existing methods in the literature. Our study is motivated from the study of Hadid *et al.* [3]. They found that at least 80% of the textures in natural images are dominated by uniform patterns, hence suggesting that the mapping pattern 'u2' is the most reliable in capturing the characteristics of natural images. We are interested to know as to

© Springer International Publishing AG 2018
Y.-W. Chen et al. (eds.), *Innovation in Medicine and Healthcare 2017*, Smart Innovation,
Systems and Technologies 71, DOI 10.1007/978-3-319-59397-5_31

whether their findings are the same in mammogram breast density classification problems. Therefore, we study the other mapping patterns ('*ri*' and '*riu2*'). To our knowledge, none of the existing studies have investigated the use of Local Ternary Patterns for breast density classification in mammography.

2   Methodology

We separate the breast region from the background and pectoral muscle, and extract only the fibroglandular disk area (FGD_{roi}). Subsequently, we use a simple median filter for noise reduction and extract Local Ternary Patterns (LTPs) to capture the micro-structure information of FGD_{roi} (see Fig. 1). Finally, we train a SVM classifier and used the model to test unseen cases.

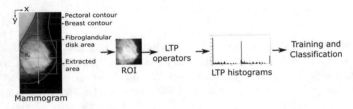

Fig. 1. An overview of the feature extraction on FGD_{roi}.

2.1   Local Ternary Patterns

The LTP thresholds the neighbouring pixels into three values -1, 0 and 1 based on the threshold value (k) set by the user. It has the following parameters: θ (orientation), k (threshold value set by the user), R (radius) and P (number of neighbours). The LTP decimal value of a pixel (i, j) is given by:

$$LTP_{(P,R)}^{pattern}(i,j) = \sum_{p=0}^{(P-1)} s_{pattern}(g_p)2^p \tag{1}$$

where R is the circle radius, P is the number of neighbours, k is the threshold constant, g_c is the grey level value of the center pixel, p is the neighbouring pixel, g_p is the grey level value of the pth neighbour, and $pattern \in \{upper, lower\}$. Once the LTP code is generated, it is split into two binary patterns (upper and lower patterns) by considering its positive, zero and negative components, using the following conditions

$$s_{upper}(p) = \begin{cases} 1, & \text{if } s(p) > 0 \\ 0, & \text{if } s(p) \leq 0 \end{cases} \tag{2}$$

$$s_{lower}(p) = \begin{cases} 1, & \text{if } s(p) < 0 \\ 0, & \text{if } s(p) \geq 0 \end{cases} \tag{3}$$

The LTP code can be generated using the following conditions

$$s(p) = \begin{cases} -1, & \text{if } p < g_c - k \\ 0, & \text{if } p \geq g_c - k \text{ and } p \leq g_c + k \\ 1, & \text{if } p > g_c + k \end{cases} \quad (4)$$

where $s(p)$ is the pth neighbour containing the LTP code value. Figure 2 shows an illustration of different LTP patterns namely '$u2$', 'ri' and '$riu2$'. Note that the black dots are neighbours with higher values than its central pixels (red dot).

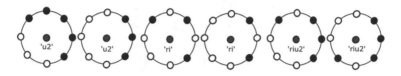

Fig. 2. An illustration of different LTP patterns with $P = 8$ and $R = 3$.

3   Experimental Results

The study was conducted based on the well known Mammographic Image Analysis Society (MIAS) database [2] which consists of 322 mammograms of 161 women. Each image contains BIRADS information (e.g. BIRADS class I, II, III or IV) provided by an expert radiologist (class I and II and class III and IV are considered as low and high risk, respectively). A stratified ten runs 10-fold cross validation scheme was employed, where the patients are randomly split into 90% for training and 10% for testing and repeated 100 times.

Table 1 presents the quantitative results of two-class classification when using different mapping patterns with different parameters values. Since this paper presents our preliminary investigation, at this stage we have set $k = 4$ and $\theta = 0$ for all experiments. The results indicate that the LTP operators produce more discriminative features when the mapping pattern '$riu2$' is used to model the appearance of the FGD_{roi}. The mapping pattern 'ri' produced the second best result on average across different parameter settings whereas the pattern '$u2$' is slightly below the performance of the other patterns. Overall, we would like to highlight the following findings: (a) the 'ri' mapping pattern performed better at smaller R compared to both '$u2$' and '$riu2$', (b) using a medium size of R tends to produce higher accuracies for both '$u2$' and '$riu2$', (c) the '$riu2$' mapping pattern produced consistent results when a larger number of neighbours (P) is used and (d) all mapping patterns are quite dependent on the parameter values.

Table 1. Quantitative results for two-class classification problem using different parameters and mapping patterns. We set $k = 4$ and $\theta = 0$ for all experiments

P	R	Accuracy(%)		
		'$u2$'	'ri'	'$riu2$'
8	2	84.56	87.75	84.71
12	3	84.78	86.84	88.59
6	3	87.31	**89.34**	86.00
6	5	87.56	88.81	88.78
16	5	**89.03**	85.45	89.65
12	5	88.03	84.59	89.65
18	7	88.68	86.23	**90.84**
12	7	87.68	85.18	90.25
8	7	88.93	87.93	89.31
18	9	87.53	86.68	89.53
14	9	86.53	85.00	89.56
10	9	85.84	89.21	90.31
8	9	85.56	89.15	89.62

4   Conclusions

Our experimental results indicate that in the mammogram breast density classification problem, using the '$riu2$' mapping pattern is more suitable than using 'ri' and '$u2$'. Our findings are different from the study of Hadid et al. [3] due to differences in datasets (e.g. their dataset contains images with distinctive and visible texture patterns whereas most breast regions contain blurred and diffused texture patterns which make it more difficult to model the appearance). In future work, we are interested in combining the LTP^{riu2} features with other texture descriptors and investigate other parameters such as k and θ.

Acknowledgments. This research was undertaken as part of the Decision Support and Information Management System for Breast Cancer (DESIREE) project. The project has received funding from the European Union's Horizon 2020 research and innovation programme under grant agreement No 690238.

References

1. Oliveira, A., Fonseca-Moutinho, J.A., Pereira, M., Freire, M.M.: A survey of the methods used to classify breast density in mammograms and ultrasound images. http://www.di.ubi.pt/~mario/Angela1.pdf. Accessed 3 Feb 2017
2. Suckling, J., et al.: The mammographic image analysis society digital mammogram database. In: Proceedings Excerpta Medica International Congress Series, pp. 375–378 (1994)
3. Hadid, A., Pietikainen, M.K., Zhao, G., Ahonen, T.: Computer Vision Using Local Binary Patterns, pp. 13–47. Springer, London (2011)

Author Index

Printed in the United States
By Bookmasters